I0139207

# A Teacher Prepares

## From Brain to Science Literacy

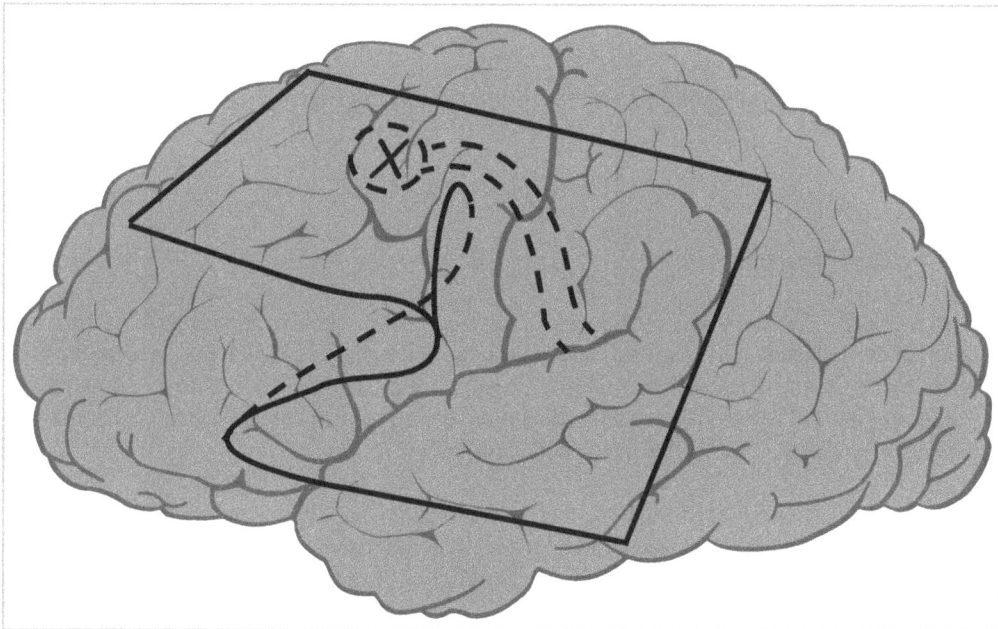

PETER R. BERGETHON, M.D.

SYMMETRY LEARNING PRESS

Symmetry Learning Press
Skytop, PA 18357
www.invariantresearch.com

SYMMETRY | LEARNING

PRESS

Symmetry Learning Press is wholly owned by Invariant Research Limited

© 2023 Invariant Research Limited

All rights reserved. This work may not be translated or copied in whole or in part without the written permission of the publisher (Invariant Research Limited, Skytop, PA, 18357 USA), except for brief excerpts in connection with reviews and scholarly analysis. Use in connection with any form of information storage and retrieval, electronic adaptation, computer software, or by similar or dissimilar methodology now known or hereafter developed is forbidden.

The use of general descriptive names, trade names, trademarks, etc., in this publication, even if the former are not especially identified, is not to be taken as a sign that such names, as understood by the Trade Marks and Merchandise Marks Act, may accordingly be used freely by anyone.

ed

9 8 7 6 5 4 3 2 1

ISBN 978-1-58447-101-1 A Teacher Prepares: From Brain to Science Literacy

# Contents

# Unit 5                      The Graphical Analyzers

# From the Creator of SymmetryScience™

When you marvel at the magnificence of the world in which we live, including your children, do you wonder at how it all works?  How do we understand it all? - The answer - by being scientifically literate.

For over four decades I have been privileged to examine the mind at the bedside as a neurologist and explored the physics of thinking in the laboratory as a biophysicist and cognitive neuroscientist.  Over these many years I have also been an educator of medical and graduate students, night-school adults, continuing  medical education health professionals, and children learning science and nature in school or at camp from elementary to middle school ages.  All of my students could recall facts but often struggled to apply these facts to problems of interest, and they had no systematic way to think scientifically.  How could our top students not understand the clarity that scientific thinking makes possible?  I became passionate to develop tools that could educate my students and to build successful interdisciplinary research work groups.

My neuroscience cognitive training from the rich experience and traditions of the great Neurological Unit at Boston City Hospital gave me deep insight into both the nature and an approach to this problem.

- As a neurologist and cognitive neuroscientist I knew the brain worked by extracting and mapping the features of the information it sensed - it built internal *models*. Understanding this has been one of the great advances in neuroscience in the last half of the twentieth century. While remarkable for its power, the brain makes cognitive errors that are intrinsic. These errors are uncovered in disease and often seen in natural human behavior.

- As a professional scientist I used systems thinking and dynamic (changing with time) modeling everyday. These methods were required to discover the "how and why" explanations of the complex biomedical systems we studied. Everyday our observable and explanatory models were tested in experimental models.  I was frequently reminded  that because time-dependent systems are always complex, we often oversimplify to solve them. As a result, we forget our assumptions and make understandable but serious errors. These are the errors intrinsic to the brain.

- As a teacher of biophysical chemistry, protein biochemistry, cognitive neuroscience, and medicine, I saw these same intrinsic brain errors in the way my students approached their studies. In addition, teachers often taught in ways that made these errors worse!

As a result of these observations, I discovered a path and a solution to the science education problem. **The key teaching insight was to foundationally teach the powerful scientific tools of systems analysis and  modeling as the language of science, especially**

**interdisciplinary study and research. This was the element that I had found lacking in my students**.

I developed and codified the constructs and tools described in this book:

- *The Progression of Inquiry* - a simple way to describe and coordinate the way we teach the scientific method with what scientists actually do in practice.

- *The Cybernetic Sequence* - a useful construct to codify model-making in the brain (developed with my colleague Dr. Alexander Woodcock).

- *The Language of Patterns* - a practical language that could be taught to all students and adults that would quickly provide competence in systems analysis and model-making.

- *The Cycle of Pedagogy* - a simple description of how models are used by the brain to learn, by humanity to generate disciplinary knowledge, and could be advanced into teaching through the practice of pedagogy.

- *The Assessment Framework Design Method* - a method to practically develop curriculum and to teach science, based on best cognitive neuroscience practices - the embodiment of the Cycle of Pedagogy. Developed with Dr. Mark Moss and colleagues at Boston University, these tools formed the basis of the successful Vesalius Program.

The SymmetryScience™ curriculum had its beginnings in the 1990s while I was a professor teaching medical and graduate students. The teaching tools listed above were being incorporated in my classroom teaching and in several biophysical chemistry textbooks that I authored. At that time my daughter had entered primary school, and I began working with teachers who were struggling to instill science literacy in their students. I was motivated to develop a curriculum of science study based on the idea that science could be more universally understood if its intrinsic language was explicitly taught early in the educational process, much as we teach children to read and write. These were the beginnings of the SymmetryScience™ program.

The concepts at the heart of the SymmetryScience™ program development recognize the cognitive neuroscience parallels between science and written language. Before the school experience begins, the brain is busy learning on its own. Children begin school with a huge vocabulary and a knowledge of the syntax of spoken language; then we teach them the code of reading and writing. Language acquisition is natural but reading and writing are not. They require instruction. Over the same time, the brain is sampling the natural world and constructing experiences into a view of the world; the code to translate this natural inquisitiveness into scientific thinking is the basis of what SymmetryScience™ teaches. But the brain is not intrinsically scientific nor critical, rather it is by nature magical. To prevent the intrinsic tendencies of the mind to reach unsupported conclusions, I developed a unified, structural approach to teaching that accounts for how the "brain learns". This approach utilizes a language of critical analysis that is central to science learning; this is the "Language of Patterns".

Using the Language of Patterns, I designed an educational program that incorporates laboratory and activity-based explorations for students from kindergarten through Grade 8, coordinating with each stage of early brain development, thus providing a unified, structured approach to knowledge construction that is built on the Cycle of Pedagogy and implements the Assessment Framework Design Model; it has become known as the SymmetryScience™ method, a method of structured discovery.

The philosophical core of the SymmetryScience™ curriculum is based on three fundamental principles:

· First, science is a way of looking at the world. Properly taught, science simplifies our view of the world. With mastery of a unifying language and a core of fundamental concepts and strategies, the limitless diversity of the universe becomes manageable and open to everyone.

· Active learning is work! This is true for the learner as well as the educator-mentor. The structured discovery design of SymmetryScience™ method coordinates how the brain incorporates new learning to provide support context for the addition of the next piece of content. The goal of making students successful is accomplished by encouraging habits of rigor and skeptical inquiry.

· The SymmetryScience™ program recognizes that teachers (whether parent, professional educator, community member, or peer) are a central figure in the learning of every student. While ultimately every educator strives to make the student independent, the process of achieving this independence needs to be supported through materials and adult learning opportunities. Parents and educators who are inspired to learn, inspire students to learn! All of our materials are designed with this in mind.

At the turn of the 21st century the SymmetryScience™ Kindergarten through Grade 8 curriculum was being piloted in almost 20 public and private schools in Massachusetts. The SymmetryScience™ approach has been successfully field tested and extended across teacher colleges and medical schools. The National Institute of Health (NIH) supported its programs for teaching middle school students in the inner city and rural America. The core of SymmetryScience™ has been used to develop training for first responders in emergency management programs across the United States, and it has informed work on the stewardship of disciplinary teaching with the Carnegie Foundation for the Advancement of Teaching. It has been used to develop educational programs in professional societies, including the American Academy of Neurology. In addition to these educational successes over the last 20 years, my research teams in bio-medical research and industry successfully applied these tools to build professional research groups that remain productive and innovative in industry today.

When I retired from professional academic and industrial science, our nation was recovering from the COVID-19 pandemic. There has never been a more powerful example of why we need to re-engage the challenge of making a scientifically literate public. I decided to revise and re-issue the original SymmetryScience™ curriculum for the broad community, because these are tools that have been proven to develop scientific literacy for everyone.

This book provides a window for the parent or teacher into the educational neuroscience foundations of SymmeryScience™. These tools bring success whether you are a parent-citizen, a first provider, or a professional scientist or engineer. Most importantly, this approach will give our children and the generations to come the power to appreciate, be stewards of, and wonder at the magnificent world in which they will live.

Please enjoy the vistas that SymmetryScience™ will open for you and your family. The SymmetryScience™ program does not just teach science, it uses science to teach; it teaches that Science is a way of looking at the world.


Peter R. Bergethon, M.D.

Creator, SymmetryScience™

# Preface

With the vast quantity of information available today, the greatest challenge in education is to harmonize the following three strands:

- How the brain processes information and yields knowledge;
- How the endeavor of gathering information has led to a knowledge of our universe;
- How to structure a coherent learning program for all citizens, especially our children.

One of the most important outcomes of an effective educational system is the ability to use critical thinking skills effectively. These skills are the language of critical analysis and model-building. Like any language, these skills are best taught starting in the early learning period of a child's development. Furthermore, the learning of these skills must be coordinated with the capacities of the developing brain.

The SymmetryScience™ method uses a systems-modeling approach to effectively teach scientific and critical thinking skills. The cornerstones of this method of teaching are the constructs of the Cycle of Pedagogy, the Progression of Inquiry, and the Language of Patterns. The Cycle of Pedagogy is the framework of the SymmetryScience™ method; it ties the developing student's brain and mind to the  system of knowledge that is to be taught and learned. Models are a central concept in the scientific and critical analytical approach to understanding the world around us. The Progression of Inquiry provides the structure for the model of science. SymmetryScience™ uses the Language of Patterns as the critical thinking language in which scientific knowledge and inquiry can be learned and expressed. The Language of Patterns is applied using the Graphical Analyzer System. The Graphical Analyzers are a practical series of knowledge organizers that guide the student to progressively learn science inquiry while teaching the Language of Patterns. Using these constructs and system-modeling tools, SymmetryScience™ curriculum has been very successful for learning science and developing successful critical thinking skills. In practice, the Graphical Analyzers are intended to structure how all students, from their earliest stages of learning to advanced study in science, using observation, description, measurement, modeling, and experimental design to learn and practice science inquiry. Using the Analyzers continuously and progressively throughout a student's schooling (from kindergarten through high school and into graduate learning), the SymmetryScience™ curriculum guides the learner to practice the common threads of scientific inquiry and construct knowledge at a developmentally appropriate level.

| SYMMETRYSCIENCE AND THE PROGRESSION OF INQUIRY | | |
|---|---|---|
| TYPICAL GRADE LEVEL | SCIENTIFIC/CRITICAL THINKING | STEP OF SCIENTIFIC INQUIRY |
| Kindergarten | Observation | Exploring the parts or elements of a system, the properties of these elements, and the arrangements of system elements into patterns. Increasing ability to pay attention - set maintenance. |
| Grade 1 | Description | Describing and defining system elements, properties, arrangements, and background space. Classifying by characteristics. |
| Grade 2 | Measuring and Description | Measuring elements, properties, arrangements, and background space and learning about cycles and processes of change. |
| Grade 3 | Descriptive Models / Cause and Effect | Developing skepticism of observation and considering cause and effect. |
| Grade 4 | Inferring Observables into Descriptive Models | Beginning to draw general conclusions from diverse information. Emergence of abstract thinking. |
| Grade 5 | Modeling and Verification | Proposing simple linkages of cause and effect. Increased ability in set-shifting/cognitive flexibility. |
| Grade 6 | Testing Explanatory Models with Verification | Constructing models of cause and effect. Forming hypotheses and testing for verification. |
| Grade 7 | Experimental Models and Design | Testing a hypothesis of a proposed model of cause and effect by experiment. Designing and using controls and variables to make conclusions. |
| Grade 8 and Above | Experimental Models and Design | Inferring unobserved properties from observed properties (observables). |

With practice and structure, the student will achieve a core competence and natural fluency in using the scientific method to explore and understand the world.

Interdisciplinary thinking is a key outcome of the SymmetryScience™ program. The SymmetryScience™ method encourages the use of the Graphical Analyzers across all fields of study in order to emphasize and widen a student's opportunity to practice the scientific method and to understand the general value of the method of scientific and critical inquiry in all fields of interest and all walks of life. This important aspect of the SymmetryScience approach encourages the students to use the scientific method constantly in their daily life, making scientific and critical inquiry a habit of mind that extends to all areas of their interest.

Mastering anything of importance requires time, work, and inspiration; progress is made by the continuous and diligent application of all three. Wisdom or understanding comes from the application of knowledge, which relies upon information. The scientific method of inquiry is the tool that must be mastered to move freely from information to understanding. As each child grows in education and develops in mind, there is a process guided by the Cycle of Pedagogy that hones this tool of understanding and knowledge. The Language of Patterns, through the Graphical Analyzer System, is a map on this "road of inquiry". This book is designed to prepare teachers, parents and science professionals involved in science education to teach the SymmetryScience™ program.

# Unit One:

# A Framework for Teaching Science

# CHAPTER 1

# Models: The Foundation of the SymmetryScience™ Teaching Method

## The Age of Understanding

The beginning of the twenty-first century heralded the turn from one remarkable era into one that is more complex and even more exciting. This was the turn from the Information Era to the Age of Knowledge.

**Knowledge** is the effective and meaningful use of information. The construction of knowledge from information requires an analytical process to knit the information into useful, information-rich models that can be used in our everyday lives. The amount of information and the rate at which it is created, dispensed, and revised is mind-boggling. Our success depends on our capacity to acquire and use this torrent of information in effective and meaningful ways.

Once we become knowledge users, the ultimate goal of applying knowledge for the creative or wise solution of problems or for creating novel things lies within our grasp. We call this step from knowledge to application **understanding**, **wisdom** or **creativity**. Each of these terms implies the application of knowledge but in somewhat different contexts.

Humans are uniquely suited to being innovative, creative, and wise. Why? Because we have brains that are inquisitive, imaginative, and highly capable of learning both from the past and in the present. Humans are unique among living things in their ability to imagine futures different from their past or present. This application of knowledge is a unique capacity of the human mind, and it is directly derivative from the unique biology of the human brain. These two properties of being imaginative and being able to learn are key characteristics of the human mind. A key goal of education is to enable the development of this human mind, enabling it with the critical thinking skills that are required to succeed in a modern society. How do we accomplish this in the construct of a teaching environment?

## Teaching Framework Tools

The SymmetryScience™ method has created a few simple constructs to provide a robust framework for the role of teaching and educating people at all ages:

- The Cycle of Pedagogy
- The Progression of Inquiry
- The Cybernetic Sequence

Their application and understanding is aided by several tools:

- Modeling and Systems Analysis
- The SymmetryScience™ Language of Patterns
- The SymmetryScience™ Assessment Design Framework Model

Each of these will be defined in this chapter. How they can be used to empower the teaching of scientific and critical thinking will be more deeply explored in later chapters.

## The Cycle of Pedagogy

The framework of the SymmetryScience™ method is the Cycle of Pedagogy (FIGURE 1). This construct ties the developing student's brain and mind to the system of knowledge that is to be taught and learned. The arrangement of teaching tools and the process of enhancing learning through the use of these tools is called pedagogy. The word 'pedagogy' accurately refers to teaching children (from the Greek *paidagogos* for a child's tutor).

FIGURE 1.  The Cycle of Pedagogy connects the brain to knowledge and learning.

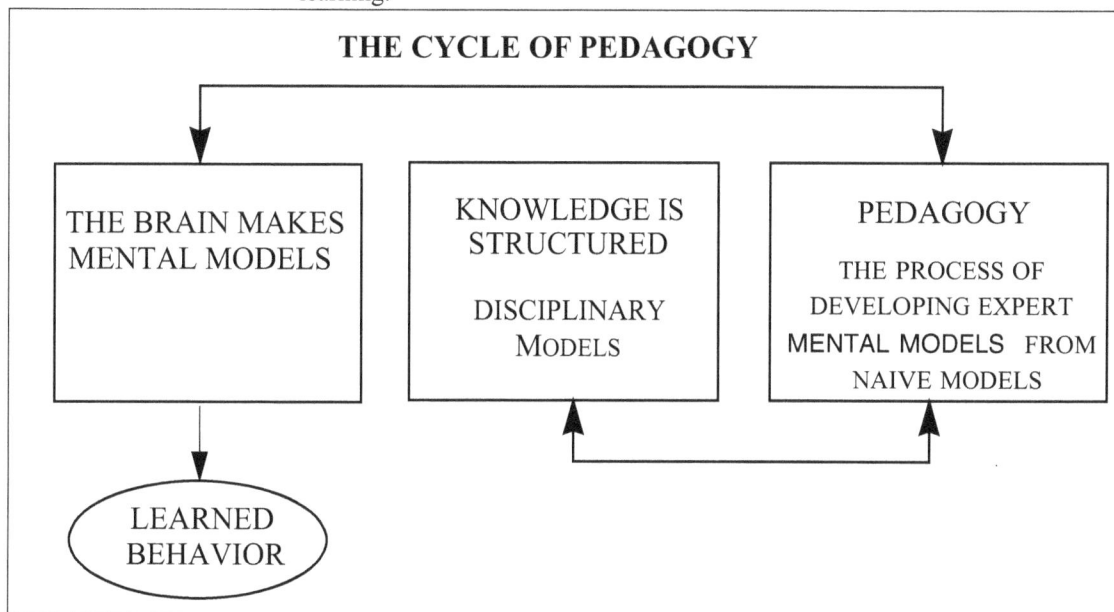

## THE CYCLE OF PEDAGOGY

| THE BRAIN MAKES MENTAL MODELS | KNOWLEDGE IS STRUCTURED<br><br>DISCIPLINARY MODELS | PEDAGOGY<br><br>THE PROCESS OF DEVELOPING EXPERT MENTAL MODELS  FROM NAIVE MODELS |

LEARNED BEHAVIOR

**Mental Models**

The Cycle of Pedagogy  states that the brain works by developing internal models.  The brain operates on these models, generating outputs or behaviors. As we will explore in Unit Two, the human brain is the learning machine of the body. However, for the first 25 years of life the brain is also developing and growing biologically. As a consequence, what and how the brain can learn at different ages depends very much on its biological maturity, which is correlated with chronological age. Although all brains share the same fundamental processes that lead to language, thought, memory, perception, and motivation, each brain is differentiated by being gifted or challenged in the variety of ways that information can be processed to produce output behaviors or learning.  After the brain matures, adults continue to learn throughout their lifetimes, although the brain shows signs of aging by the fifth decade; psychological and motivational states become more important in adult learning.

**Naive to Expert Models**

Teaching, or pedagogy as it is called (especially when teaching children) seeks to take a knowledge model and present it to the developing brain. The intention of education is to guide the brain to build a progressively more sophisticated internal model, moving from a naive toward a more expert state. A key to advancing the internal mental model from simplistic to advanced depends upon what the student already knows. Successful pedagogical techniques engage the student, define what they do or don't know, and modify the knowledge in a brain development -consistent fashion. Unit Three will explore this more fully.

The Cycle of Pedagogy is summarized by:

*"The brain learns the knowledge of a discipline when the subject is framed so the brain can learn best."*

The focus of the teacher is on helping brains learn and become educated. The focus of the SymmetryScience ™ curriculum is provide the tools to accomplish this task.

## Knowledge

The knowledge of any discipline is structured systematically into an information set that includes content, concepts, context, and processes that represent the disciplinary model of how the world is regarded. Scientific knowledge is extensive in terms of content, as the time frame of scientific discovery is often quite fast; science is a discipline that is concerned with finding objective truth through constant experimentation. Thus a large part of scientific knowledge is the continuous input of new information that must be incorporated into the current models of knowledge. Unit Four will explore the structure of knowledge systems more fully.

## Defining Learning, Education and Schooling

*"I never let my schooling interfere with my education."*

Mark Twain

"See one, do one, teach one."

The apprenticeship model of medical education

Notice that in these familiar expressions the process of learning is not mentioned. This is because most of the time the process of learning is assumed.

'Learning', 'Education', 'Schooling' are all terms that are familiar and commonly used interchangeably. They are related but distinct in the following way:

"Students will be **educated** in the discipline of science by

**learning** scientific thinking in the home **school**".

Here are useful (and the dictionary!) definitions.

## Learning

**Learning is a biological phenomenon.**

**Learning** is the modification of behavior through training, practice, or experience. This is a psychological definition that implicitly recognizes that it is the brain that is learning, thus causing the changed behavior. As will be discussed, modern cognitive neuroscience allows us to understand how the brain is modified such that the behavior of the human being is changed. Effective learning is characterized by three properties:

- a change in behavior due to learning intervention.

- durability of the change effect.

- transferability of the changed behavior to more general conditions.

## Education

**Education is the acquisition of previous knowledge.**

**Education** is defined as Mark Twain used the term in the quotation above. It is the act or process of imparting or acquiring general knowledge and of developing the powers of reasoning and judgement. Education and scholarship are related. A scholar is defined as a person who is sophisticated and with a profound knowledge of a specific subject. **Scholarship** is the term applied to the process and output having the qualities, skills, and attainments of a scholar. Scholars are subject experts who are masters of a knowledge domain. Education is the process of advancing a learner from a naive knowledge state to an expert or even master knowledge state. Practically, the educational states of an individual can be characterized in the following order: naive, apprentice. journeyman, expert, master.

## Schools and Schooling

**Schools are places for schooling.**

The instruction, education, or training (especially when received in a school) is called **schooling**. Schools are defined as places or institutions where schooling occurs. Typically, schools are defined as the places - building, room, or institutions - where education is imparted. Schools are places that often have a philosophical purpose and therefore refer to situations and places where instruction and indoctrination occur. It is clear that this is the differentiation to which Mark Twain was referring in his quote.

## The Progression of Inquiry

The Progression of Inquiry (FIGURE 2) shows how knowledge is acquired. The Progression of Inquiry is a process of sequential models that proceed from a description of observations to explanations of this observed system.

FIGURE 2. The Progression of Inquiry represents the method of knowledge discovery through sequential model-making.

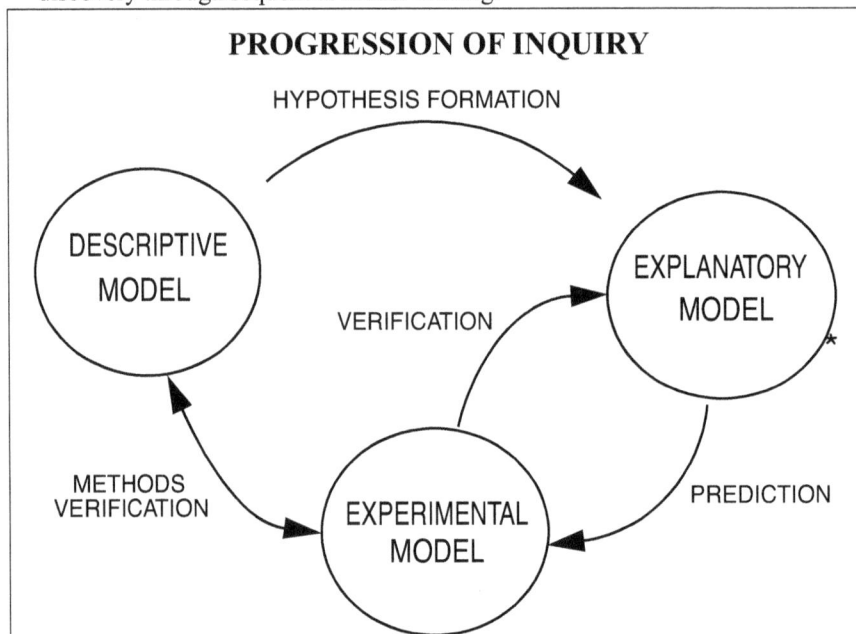

Non-scientific knowledge can be represented by these descriptive and explanatory models, but scientific knowledge can only be obtained when an experimental model is included. The experimental model and testing then provides verification and validation of the descriptive and explanatory models.

The knowledge captured by this Progression of Inquiry becomes the disciplinary models represented by the middle box of the Cycle of Pedagogy in FIGURE 1. The construct of the Progression of Inquiry shown here captures the modern scientific method, which uses experimental models to confirm the observations captured in descriptive models and causal explanations proposed in explanatory models.

The Progression of Inquiry construct depends on models and modeling. The Cycle of Pedagogy is framed in models as well. Models are a central concept in the scientific and critical analytical approach to understanding the world around us.

## The Cybernetic Sequence

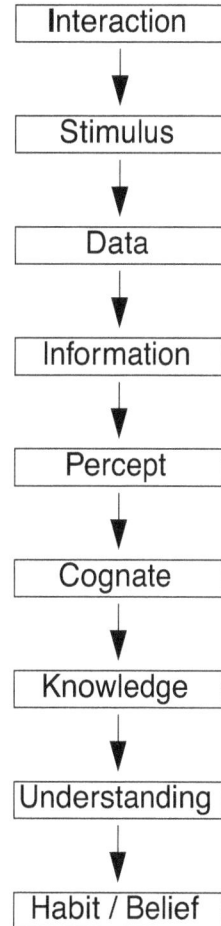

.The real world is detailed and complicated. Every brain is challenged with such richness. The solution is to simplify - usefully. These simplifications are models. A systematic exploration and simplification of our world gives rise to the models that our brain uses to represent reality. Nature uses nervous systems such as the human brain to perform this systemic abstraction and model building.

**THE CYBERNETIC SEQUENCE**

Interaction
↓
Stimulus
↓
Data
↓
Information
↓
Percept
↓
Cognate
↓
Knowledge
↓
Understanding
↓
Habit / Belief

**FIGURE 3.** The simplified cybernetic sequence.

The construct of the Cybernetic Sequence provides a way to picture how billions of details in the real world are processed by the brain from overwhelming representation into a simpler but more meaningful model of that real world. In its simplest form the Cybernetic Sequence (FIGURE 3) shows how each human interacts with the world around them through their senses, and uses sequential modeling to extract information and knowledge from this stimulus, leading to outputs that are habitual and often dogmatic.

The world we live in provides many physical interactions with each of us as biological beings. Whether we are considering the world through our primary senses (eyes, ears, nose, etc.) or any machine we use to extend our senses (thermometers, telescopes, medical lab tests or engineering tools), the interactions between the sensing instrument and the real world is all we can detect. The Cybernetic Sequence reminds us that what we detect and how we apply or use that interaction depends on the internal models our brains use to interpret the interaction.

No sensor can capture all of the natural physical details of the world, so the output or stimulus signal from the sensor is a simplified approximation (a sampled model) of the interaction experience. This sensor output is called **data**. Data are the list of what we usually think of as primary and complete, but it is not. It is already simplified by the sensors. Data become organized and simplified and become **information**. The information is actually picked over data. It is given some meaning or salience when the brain pays attention to it. If the information has some meaning to the brain it is likely to be incorporated into a knowledge structure (or model). This stimulus/interaction then develops meaning and can be acted upon, showing **understanding**.

A simple example is the best way to appreciate the cybernetic sequence in the brain (FIGURE 4).

A  **Interaction** from the green light with the retina generates...

B  **Stimulus** in the photoreceptors that send signals as...

C  **Data** through the optic nerve to the brain is optical...

D  **Information** that is processed to generate the neural...

E  **Perception** of green light which is brought to mind as a....

F  **Cognate** that a light called green is in the field of view.

G  **Knowledge** that "green means go" - "red means no go" leads to...

H  **Understanding** that the car should proceed forward.

I  **Habitual action** of driving produces a coherent behavior

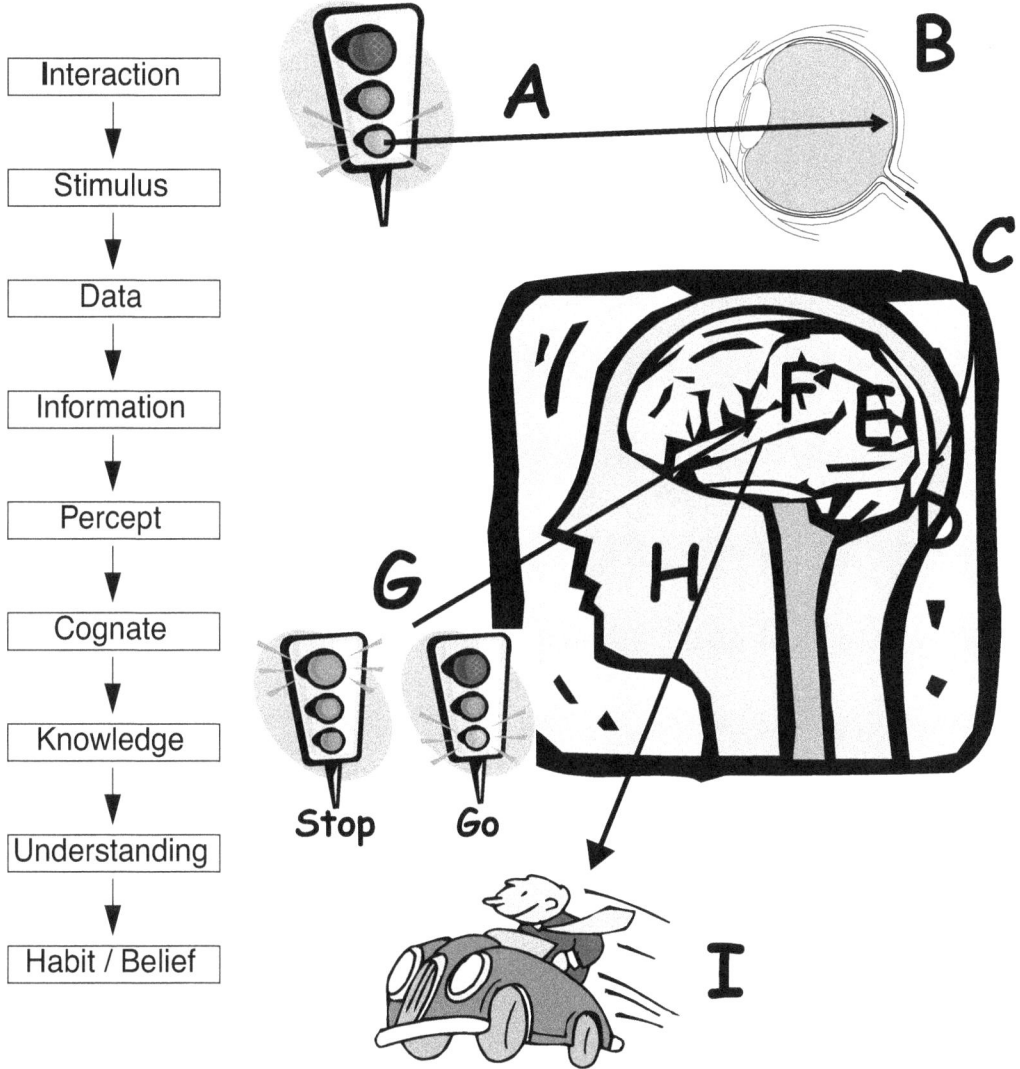

FIGURE 4. The Cybernetic Sequence in Action

## Summary

The common theme in the Cycle of Pedagogy, Progression of Inquiry, and Cybernetic Sequence is that the systems comprising the world are described using models. Models are simplifications of our real world. The framework for teaching, learning, discovery, and creativity all share the central concept of modeling. The brain knows its world by systematically reducing it to models. Human knowledge is structured into systems models of content, concepts, context, and processes. Teaching is a matter of advancing internal mental models from naive to expert levels.

The next step is to gain familiarity with the tools of systems thinking and modeling. A language called the Language of Patterns can be used to describe models and systems. Finally, the paradigms for organizing systems for teaching and engineering new things are natural extensions of that system modeling.

# CHAPTER 2

# Systems and Models in Learning

## Modeling

The central idea in the Cycle of Pedagogy is a recognition of the central role played by **modeling** in how the brain learns. Modeling is the process of representing something (called a system) using a partial accounting of its features and properties. Modeling is a *simplifying* process. This means that the model is an abstraction (a simplification) of the thing under investigation. A model, even though it is a partial representation of the real world, is acted on and treated as if it is the original. A "good" model provides a satisfactory representation *for the purposes under consideration.*

Often modeling is confusing at the start. This is because sub-models are used to build up a larger model of the particular thing of interest. The result is that there will be a series of abstract representations and symbols used to represent a complicated system. It takes tools and practice to keep track of the parts of a system and its series of representations. SymmetryScience™ developed the Language of Patterns as a tool to identify and keep track of these parts.

A model is a description of a system that *includes identifying the parts, relationships, context, and characteristics of that system* under observation. The Language of Patterns and the graphical organizers that are explained later in this book will help you learn how to make these descriptions.

## Modeling and Building Knowledge

Modeling is the coordinating link between observation and understanding. Critical analysis and critical thinking are enabled by the key skill of modeling systems under consideration. While modeling is central to the practice of the scientific method, models are also how the brain represents the world that it regards. The brain's models are not intrinsically scientific, as will be seen. Our brains extract certain features of the world; the brain uses these pattern features to construct a series of interrelated internal models at several levels:

- **Descriptive or representational models** - describe what parts there are and how they go together.

- **Explanatory models** - ascribe cause and effect relationships between the parts in the system observed.

- **Predictive models** - used by the brain to predict and plan for future events and actions.

These models form a progressive series moving toward increasingly abstract relationships. In the human mind, such a series of internal mental models represent the movement from "simple" experience to "expert" knowledge and understanding of our world. However, the expert knowledge does not have to be scientific knowledge in the human mind - in fact for most of human history, expert knowledge has been "received" or "a way of knowing" which are the expert systems of magic and religion.

Human brains are not naturally *scientific* model makers. The human progression of inquiry always involves description and explanation. Scientific modeling requires the addition of experimental models with strict rules of verification. The brain does not require this additional level of modeling in order to accept or use an explanation. It is important to be aware of this limitation since assumptions, beliefs, and opinion are often substituted in models. They are mistaken for more objective, verifiable truths. This can lead to substantial error in the modeling process. The scientific method uses specific steps to verify models. In the remainder of this book we will be concerned with scientific models and the scientific progression of inquiry.

## Modeling in Science

Scientists, applying modern scientific methods of inquiry, use a parallel series of model-building steps to learn about, organize, and explain events in the natural world. Scientists:

- observe phenomena and make **descriptive models** that define the features and pattern elements,

- then try to explain the phenomenon by creating an **explanatory model**. The descriptive model is re-organized into an explanatory or theoretical model. This is accomplished by posing **hypotheses** or proposed relationships between the system parts. These relationships are proposals for connecting how one part influences the others and ultimately the overall output properties of the system.

- A unique aspect of scientific thinking is that there must be a careful testing of the proposed explanatory relationships (hypotheses) by using **experimental models** to verify the explanatory model against evidence collected from the real or natural system.

As evidence accumulates to support the hypothesized relationships, a scientific theory is evolved. A **theory** is a model of current understanding that results from the interplay of observation, descriptive modeling, explanatory modeling, and experimental evidence.

These models (along with their components, assumptions, and relationships) are the "content" known as scientific knowledge and "taught" in science education. Scientists use the theoretical models as a starting point to ask questions and explore natural phenomena to a deeper degree. As new scientific information (observations and measurements) are made, the explanatory or theoretical model is either confirmed or changed to accommodate the new information. Thus the process and content of the progression of inquiry to produce scientific knowledge is intertwined in the enterprise of science.

The construction of the models in the Cycle of Pedagogy, Progression of Inquiry, and Cybernetic Sequence requires a systematic language that renders a model of reality. These systems are described by recognizing that they contain definable **elements** that can be **related** to one another and appear in the **context** of a particular background space or dimension. The interaction of the parts with one another and their context space gives rise to system **properties** that are often distinct from the individual characteristics of the components of the system. The language that captures systems information is a key tool in teaching critical thinking and science. In the SymmetryScience™ method this is has been called the "**The Language of Patterns**". It might also be called the "Language of Models". Use of this language is a key approach and tool needed for mastering the process of critical thinking.

## The Language of Patterns

The Language of Patterns is a descriptive modeling language. It is used at each step of the knowledge construction process or Progression of Inquiry. When an observation is made, the Language of Patterns expresses the system or entity under review coherently. Any object or system of objects can be fully characterized by a description of its pattern features, which include the emergent **properties** that result from the interaction of the system **elements**, **rules**, and **background** space. Using these fundamental ideas, the Language of Patterns develops the tools for the young learner for the critical thinking and analysis that underlie all knowledge construction.

This method of learning is simple yet powerful enough to advance understanding in rocket science, medicine, electronics as well as economics, and ice hockey play-making! Despite its capacity to analyze complex topics, the method is so easy to use that a kindergarten level student can use it at the start of the school year. Once mastered, the Language of Patterns can be used over and over to characterize, explore, and understand any system of interest at multiple levels.

Scientific inquiry uses the Language of Patterns at every level from observation, to model building, to hypothesis testing, to experimental verification. Critical thinking skills and the process of scientific inquiry are practiced anytime the Language of Patterns is used, anywhere, for anything. When a model is built, the Language of Patterns allows symbols, such as chemical, technical, and medical notations, as well as letters, musical notes, and mathematical equations to be assigned to each property, element, arrangement, and background space.

As the need for the logical tools of deduction and inference develop in the modeling process, the Language of Patterns is the method of choice to extract the rules and connections from the constructed model.

During the **verification process**, a hypothesis is posed which can be proven or disproven by collecting evidence. The Language of Patterns relates the observation to the model, to the hypothesis, and to this gathering and interpretation of the experimental data.

Finally, when the knowledge construction process calls for opening the discovery process to outside evaluation, the Language of Patterns provides a common, precise language of communication and discussion.

The Language of Patterns is structured just as scientific inquiry is structured. Therefore:

· Teaching the Language of Patterns is teaching science.

- Using the Language of Patterns is doing science.

- Teaching with the Language of Patterns is teaching with science.

A series of graphical organizers (the Graphical Analyzer System) assist all users to become fluent in and apply the Language of Patterns throughout their studies and work. Learning the Graphical Analyzer System is detailed in the final unit of this book.

| | |
|---|---|
| **Systems, Models and Creativity** | The world around us can be discovered through the Progression of Inquiry. The work that we imagine and wish to build can be created through the creative design process. Both discovery and creative design are sequences of Descriptive, Explanatory and Experimental modeling. The Language of Patterns as a language of systems and models gives us critical analysis tools to discover and to create. The approach maps directly to engineering design, computer programming (especially object oriented programming) and other creative processes in writing, music, theatre, cinematographic arts, and architecture. The SymmeryScience™ method has developed the Assessment Framework Design Model as its basis for effective constructivist teaching. |
| **The Assessment Framework Design Model** | The Assessment Framework Design Model (AFDM) implements the cycle of pedagogy by organizing what is to be learned systemically. It explicitly recognizes the cognitive neuroscience view that the brain utilizes internal models and that learning must update and modify these internal models. This is called a constructivist process. |
| | The AFDM is scientific in its formulation because it places measurement of the changes in the internal models as a key part of the teaching process. Explicit assessment measures: |

- Where the learner begins,

- How they are changing, and

- If the changes in the internal models are manifest in changed behaviors.

| | |
|---|---|
| **AFDM Principles** | The principles upon which the AFDM are based are: |

- Learning is a constructivist process (not a behaviorist process) and generally requires a learning cycle to implement the learning.

- Constructivism is intrinsically a modeling process and therefore requires systems analysis and design.

- Learning cycles require assessment processes to be the framework on which learning is monitored and are a critical part of the knowledge infrastructure, not an afterthought.
- The merging of systems analysis/design, cognitive neuroscience measurement, and learning cycle processes is the paradigm for the AFDM.
- The Cycle of Pedagogy is a practical guide to the interaction of the learning machine (brain), the structure of knowledge and the teaching methods used to transfer that knowledge into the learning brains (pedagogy).

**The Learning Cycle**

A learning cycle (e.g., the 5 E's) is the term used to describe the steps in a pedagogical paradigm that guides construction of internal models. A learning cycle coordinates the teaching steps with cognitive neuroscientific understanding of learning:

1.   **Engage** - focuses attention and attaches emotional salience to the learning event.

2.   **Elicit** - what is the state of the internal model already held by the learner?

3.   **Explore and Explain** - Advance new material to the student at the current level of student understanding.

4.   **Elaborate** - use reflective and critical analysis tools to refine the internal model and advance their  complexity in the direction of greater expertise.

5.   **Evaluate** - recognize the essential role of assessment to the process of reflective and responsive constructivist learning.

**Knowledge is Structured**

In the context of the AFDM, the structure of a knowledge system is described with a systems approach - the components of which are called the 3 C's + P.

1.   **Content** - maps the elements of the knowledge system.

2.   **Concept** - maps the relationship rules of the knowledge system.

3.    **Context** - maps the relationship background context to the knowledge system.

4.    **Process** - the discipline specific tools and skills used within a knowledge domain.

# CHAPTER 3

# The Character of Scientific Literacy

## Basis of Science Literacy

Science literacy is knowing how to observe, define, and describe the natural world, and code it into a systematic modeling language.

In addition to understanding the process of scientific fact finding, scientific literacy includes being familiar with the central ideas of how the natural world works.

FIGURE 5. Scientific literacy is a system of processes and content.

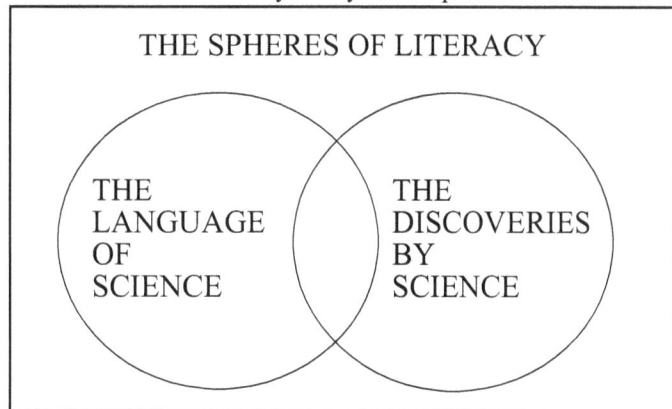

THE SPHERES OF LITERACY

THE LANGUAGE OF SCIENCE

THE DISCOVERIES BY SCIENCE

Skillful use of critical analysis and an appreciation of the principles of experimental design are the tools of a scientifically literate individual.

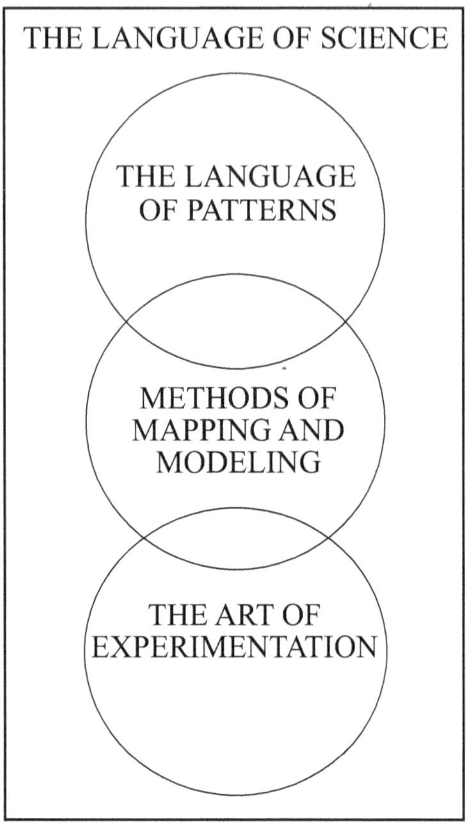

FIGURE 6. Science is a language of interacting models: description, explanation and experimental empiricism.

Literacy in science includes:

- The ability to describe and analyze systems. In SymmetryScience™ this is the **language of patterns**, which is how information is handled for description and analysis.

- The techniques that transform the pattern features from an observation onto a model system.

- The model system may be movement, graphical, written or verbal, or symbolic (mathematical, chemical/ physical notations).

- The art of experimental design.

THE LANGUAGE OF SCIENCE

THE LANGUAGE OF PATTERNS

METHODS OF MAPPING AND MODELING

THE ART OF EXPERIMENTATION

- A skeptical view of the observed and theoretical worlds.

- The tools used to explore and measure the natural world.

- The recording of the information.

- A sense of the enterprise of scientific investigation.

- The endeavor of science practiced in scientific laboratories.

- The perspectives of science.

- The scientific habits of mind.

## Categories of Science Literacy

A literacy in scientific knowledge fall under the categories of:

- Physical sciences - these include the domains of physics and chemistry

· Biological science

· Earth and space science

The content of discoveries about the natural world made through scientific investigation can be understood from the viewpoint of just several major ideas.

· **Space and time** - Everything in the natural world is contained in a background space that has height, length, and width and exists over a period of time.

· **Atomism** - The basic elements of the world

· **Interactions** - Between elements as described by energy and conservation rules

· **Structure** - How arranging the atomic elements give rise to a wide variety of structures with special properties and behaviors.

· **Change** - How structures and properties change or **evolve** over time, responding to interactions with the environment.

Each of these ideas are related.

**Space and time** provide the background in which the **atomistic elements** interact with each other according nature's rules of **interaction**. This gives rise to all of the **structures** of the natural world which have measurable properties and behaviors and interact with each other. This level of interaction leads to **change or evolution** over time of one set of structures into new and different ones.

These foundation ideas are applied in the disciplinary fields. A literacy in science includes a basic knowledge in each of these fields.

| **Physical Science** | Structure of Matter |
|---|---|

· atomic structure

· chemical elements

· chemical compounds and properties

Interactions Between Objects (Chemistry, Physics)

· forces in nature

· forces and motion

Energy Flows and Changes

· conservation of energy and matter

- statistical behavior in the physical world

- energy transformations and disorder

| | |
|---|---|
| **Biological Science** | Physical and Chemical nature of Life Processes<br><br>· molecular basis of heredity<br><br>· flow of energy and matter in living systems<br><br>The Change of Biological Structure and Diversity Through Biological Evolution |
| **Earth and Space Science** | Evolution of the Universe: Planets, Stars and Galaxies<br><br>Matter and Energy in the Earth System<br><br>Geochemical Process and Cycles in the Earth System |

## Skepticism in Modern Science

In modern science, a skeptical view is taken of what is known. A skeptical view is characterized by the continual questioning of the certainty with which something is considered "true". "How sure am I, that I actually know what I think I know" is the skeptical view. This includes a skepticism of the ability of the human to objectively make observations. The scientific methods of error-checking use experimentation and the experimental examination of a natural system. Experimental-skeptical methods include:

- **Control** experiments to be certain that the effect observed is related to the proposed cause.

- **Calibration** of measurements to be sure that an agreed upon standard is used to quantitatively compare measured effects between experiments and experimenters.

- **Context** of the system under study with reference to other systems.

- **Communication** between experimenters to confirm and criticize each others experimental and theoretical work.

- **Statistical thinking** because the universe is always randomly variable and there is always a degree of uncertainty in all measurement and observation. Variability and uncertainty must be measured and accounted for in order to be sure that an observed effect or property actually exists.

| | |
|---|---|
| **Art and Science** | Art is a human process whose aim is to explore the relationship between any pattern and the human perception of that pattern. Both science and art explore many of the same aspects of the world. In pre-modern times, art, science, technology, and religion were often facets of the same stone. |
| **Language and Critical Thinking** | In critical explorations, language plays a central role. Language is an essential tool to critical, skeptical analysis. Language provides the capability to precisely define and assign the properties that are required to abstract the patterns perceived by the brain. These properties can then be generalized. Language can be used for logical, rational analysis and also for opening channels of communication in which different people can propose, disagree, and argue with particular viewpoints. Thus, language is a fundamental tool of critical analysis. Its skillful use is necessarily highly regarded because of its capacity for abstraction, imaginative exploration, rational exposition, and communication. |

# CHAPTER 4

# The Path to Science Literacy

## What is Literacy?

Most people think about literacy as being able to read and write. Literacy is knowing the letters and how they are arranged to form the sounds and words of a spoken language. The words are then arranged to make sentences, poems, essays, stories, and dramatic works. Like reading and writing, literacy in any area is a mastery of the codes or languages used to describe the knowledge in that area.

In the modern view, literacy is both the command of the language, tools and codes of a discipline and also its ideas, content, and history. What does it mean to be literate beyond language in other areas, such as mathematics, art, music, or "the culture"?

- In mathematics, the language of literacy would be numbers, set notation, geometry, and symbolic logic.

- In art, music, civics, and physical activity, literacy is gained with the use of the languages of shape, color, materials, notes, rhythm, laws, respect, justice, movements, dance, etc.

## What is Science Literacy?

Science literacy is knowing how to observe, define and describe the natural world objectively. Science literacy requires knowing how the observations and their explanations are tested. In addition to understanding the process of scientific fact finding, scientific literacy includes being familiar with the central ideas of how the natural world works. Skillful use of critical analysis and an appreciation of the principles of experimental design and practice are the tools of a scientifically literate individual.

Why is science literacy so important to the individual? Science is the best way of exploring and knowing about the world. We make sense of our world through the lens and tools of science. Through the lens of science, the principles underlying art, music, mathematics, language and communications, health and athletics, and history are clear. Science allows us to understand the beauty, depth, and complexity of our world.

Why is science literacy important to society? Science, as a way of knowing, is the de facto standard in all aspects of our society. Scientific inquiry defines the application and use of evidence to verify observation and ideas in the marketplace, the courts, in art and literary criticism, and ideally in social and political discourse. The children of today will be the adults of tomorrow and they must have the tools to make decisions about their lives and their world. Science literacy is the key to gain access to and be included in the machinery of our civilization. Inclusion and participation are the path to democratic citizenship.

How can science accomplish all of these things? To explore this question, we must know more about what is meant by 'science' and by extension scientific knowledge.

## What is Science?

Science is a human endeavor that explores the natural world. It is the human brain that does the exploring. At its most basic level, science is a knowledge of the natural world. A knowledge of the *patterns of natural events and the resulting effects* are among the most basic interests of all people. Since the dawn of our species, survival has depended on recognizing these patterns which have included:

- day-night cycles
- the motions of the sun and moon
- the turn of the seasons
- the migration of animals
- the rhythm of human existence.

All science starts with observations of the patterns of the natural world.

Archeological evidence from cave paintings and the notching of bone and reindeer horns suggests that pre-historic humans were extremely careful in their recording of seasonal and temporal patterns. Such knowledge is acquired by simple observation.

A hunter-gather society depends on this science to:

- know where and when animals gather, feed, obtain water, or sleep

- know when and where berries, shrubs, and flowers are located and will bear fruit

- know the patterns of weather, drought, and flood so that migration ensures the survival of the society

An agricultural society also depends on this basic science so that it knows:

- when to plant

- when to reap

- when to gather and store food

Observation alone is not "modern" science. Observation is "proto-science". The human brain always tries to organize observations of its world into explanatory *models* of cause and effect. This drive is unique to the human mind and is shared by no other species.

## Correlation versus Cause

Many animals detect and act on correlations, which are perceived relationships within observed or experienced patterns. Making correlations is the basis of learning. However, correlation is not causation. Causation requires a mechanism. Humans are endowed with neural machinery that allows making this differentiation: the human brain is driven to understand cause and effect or how and why things happen.

Models of cause and effect are built on a *mode of explanation*:

Modern scientific thinking requires a particular kind of proof to provide an understanding of how cause and effect are connected.

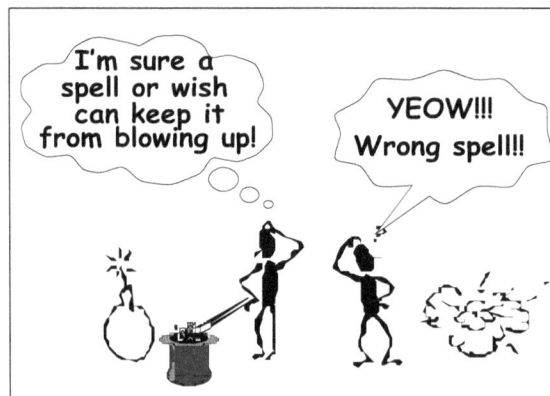

Non-scientific modes of explanation are often naturally applied to "proto scientific" observations.

The task of the brain is to gather the information (observation) to understand what it means (mode of explanation). In order to understand how the brain uses models and modes of explanation, we must understand the brain.

# Unit Two

# The Mind Makes Mental Models

CHAPTER 5

# The Progression of Inquiry and Modern Science

## Modes of Explanation

A "mode of explanation" establishes the way that cause and effect relationships explain the natural world. The human brain seeks a cause for the effects or patterns it observes in the world. Correlation can be erroneously substituted for cause, but causal models are always sought. Understanding why and how something happens is the basic goal of human curiosity.

Historically, humans and their cultures have used several modes of explanation in their attempt to understand the cause and effect patterns captured in their observations of the natural world. The unique quality of the scientific mode of explanation is its dependence on the use of evidence to test a proposed model of explanation and to use evidence to correct errors in those models.

Three principal modes of explanation seen in the history of humans are:

- received knowledge
- ways of knowing
- modern scientific skeptical empiricism
- and most recently, modern science

## Received Knowledge

A mode of explanation that attributes the cause for events to gods, magic, and mystical powers leads to mythological and theological explanations for the events discovered in the natural world. Mystical attribution is usually based on "received knowledge". The observed evidence, which may be quite accurate and detailed, is interpreted in a god-demon-magic context.

For much of human history, patterns of importance to hunter-gatherer and agricultural civilizations have been noted and used to ensure survival of the society. For most of human experience, the causality relations were attributed to supernatural gods and magical. occurrences. Thus, a "proto-science" based on observation with theological attribution has existed for most of humankind's tenure.

**FIGURE 7.** Myth and magic do not need validation.

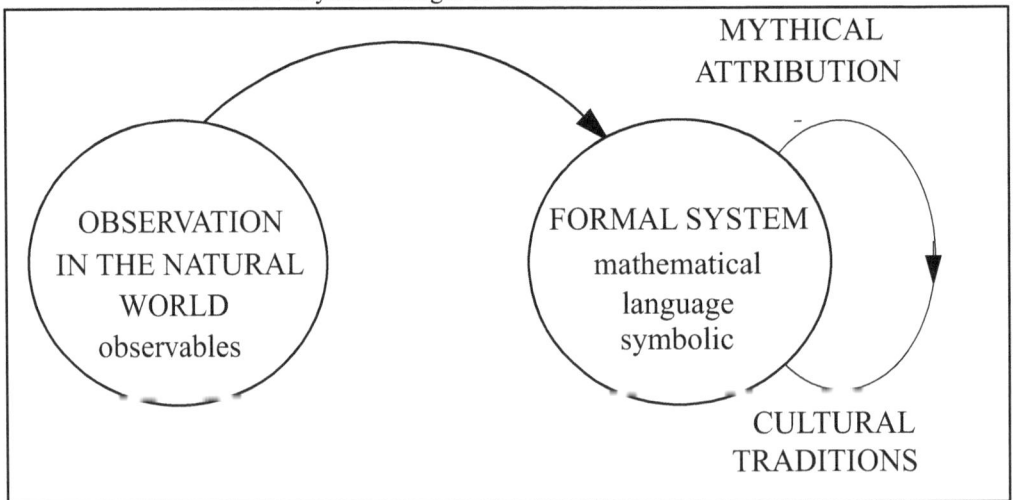

# Ways of Knowing

A way of knowing is typically derived from a frame of reference assumed to be true based on an internal feeling, identity, or experience. The viewpoint while often built upon highly logical and rationalized deductions typically derives from initial premises derived from personal experience or a personalized sense of unique access to foundational truths. Ways of knowing often result in utopian formulations.

The ancient Greeks made extremely careful observations about their natural world and developed models that explained the observations according to strictly rational - logical deductions. The starting point for these deduced models were derived from "self-evident" truths. These models of thought assume that the actions of the universe are rational according to a human-rationalized order. In the Greek (Aristotelean) view, the philosophical mind saw truth and perfection in the mind and imposed it onto the universe.

For example, the Greek view on motion was that:

- the gods who made the world are perfect

- circles and straight lines are perfect

- gods make motion

- motion must be perfect because the gods made it

- therefore, motion in the natural world is circles and straight lines

So planets move in circles and cannon balls move in straight lines.

The problem with models explained by "ways of knowing" is that all observations are forced to fit the model or they may be ignored. The underlying model can not be changed by evidence.

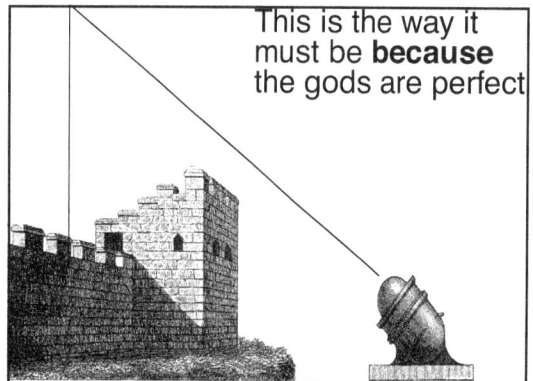

This is the way it must be **because** the gods are perfect

This mode of explanation is resistant to any change in world view because new observations can not alter the underlying models of cause and effect. For example, the idea that motion occurred in straight lines led medieval military engineers to calculate that a cannon ball would rise to a certain height and then fall straight down over a castle wall. The problem was that the cannon balls did not land where the medieval engineers expected. The Aristotelean "way of knowing" was not able to provide a means to correct the error between what was expected and what happened.

**FIGURE 8.** Special perspectives and magic are not self-correcting

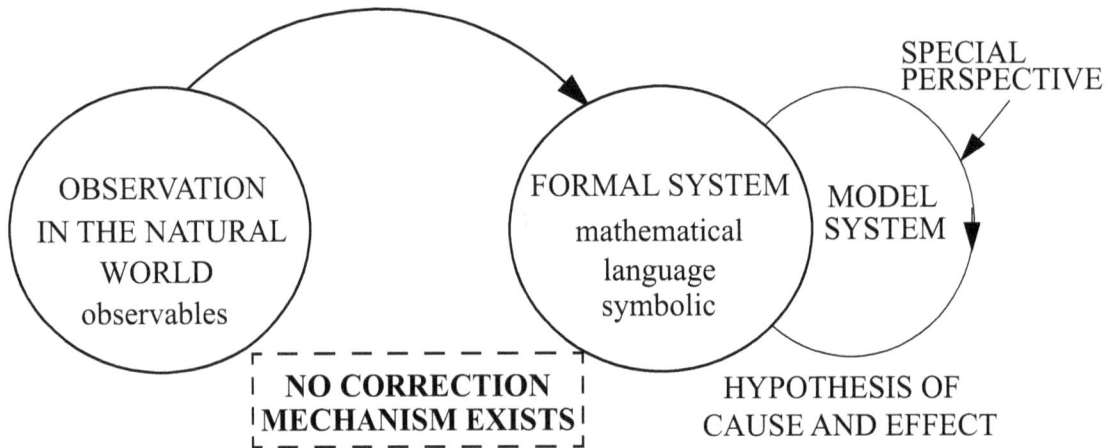

Both of the "received knowledge" and "ways of knowing" modes of explanation satisfy the brain's goals of completing patterns of cause and effect and of avoiding ambiguity. However, the particular viewpoint of these modes of explanation enhances the pre-set bias of the knowledge surface behavior of the brain. Neither has the intrinsic capacity to alter the underlying world view. These modes of explanation are therefore limited in their flexibility and capacity to expand their field of knowledge beyond a relatively restricted plane of observation. Both are like looking at the world through a fixed focus lens or in the most narrow case, closing the lens completely and considering only what is already known and accepted as the extent of relevant knowledge.

**Scientific Mode of Explanation**

"A scientist commonly professes to base his beliefs on observations, not theories....I have never come across anyone who carries this profession into practice... Observation is not sufficient...theory has an important share in determining belief."

Arthur S. Eddington, The Expanding Universe, 1933

In the Italian Renaissance, Leonardo da Vinci, who was a very good military engineer, tried to solve the problem of the "Missing Mortar Shells". Da Vinci's approach was radical for his time. He observed that when a mortar was fired, the shell followed a path that was not the one predicted by "perfect" motion. There was no straight line motion at all! Instead the shell followed the path of a "parabola". He changed his world view based on his experiments and measurements. Da Vinci's mortars began to hit their targets and the seeds of experimental modern science were planted.

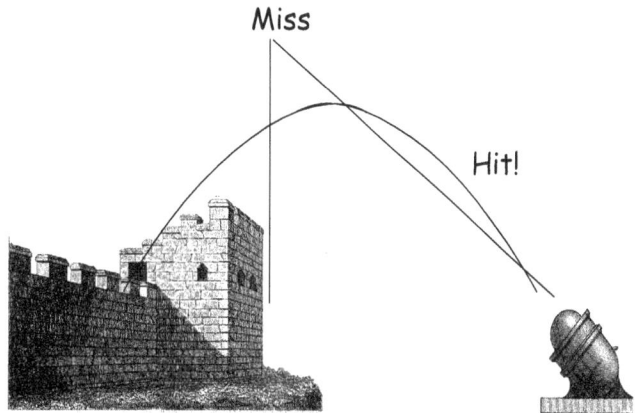

In the scientific mode of explanation, the fundamental rules of the explanatory model are discovered not from assumption or philosophical musing, but rather from careful consideration, measurement, experiment, and analysis of specific, relatively simple cases.

## The Modern Scientific Method

Observation is the first step in constructing a model. Observations are made in the natural world. The measured attributes are mapped onto a theoretical model. The model is usually made up from a "formal system" of equations, words, geometric figures. The validity of the model is tested by:

· making a prediction.

· performing experiments in the natural world to test the model.

· making experimental measurements while recognizing that the observer may influence the experiment, and measuring that influence.

· changing the model with experimental evidence.

This overall process is the modern Scientific Method.

FIGURE 9. Modern Scientific Method

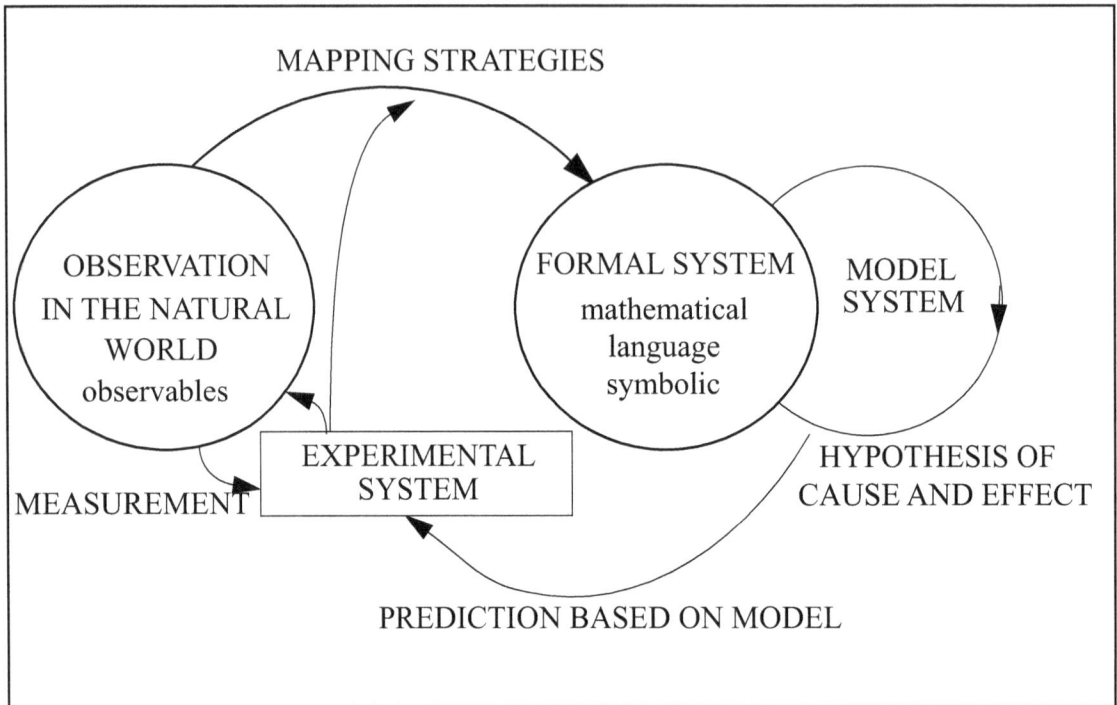

The scientific mode is distinguished by the constant modifying of the cause-effect model with experimental evidence that verifies or falsifies the ideas on which the model is constructed. The experimental model is the key validation component.

## Application of the Scientific Method

An example of how the scientific method interactively corrects a plausible but incorrect model is worth considering. We will consider the well known parental truism (Figure 10):

*"Getting chilled (a wet head on a cold day) increases the chance of catching the common cold"*... or does it?

**FIGURE 10.** Following the scientific process shows that the causal model of weather causing viral illness is unsupported by evidence.

# Start at A and follow to G

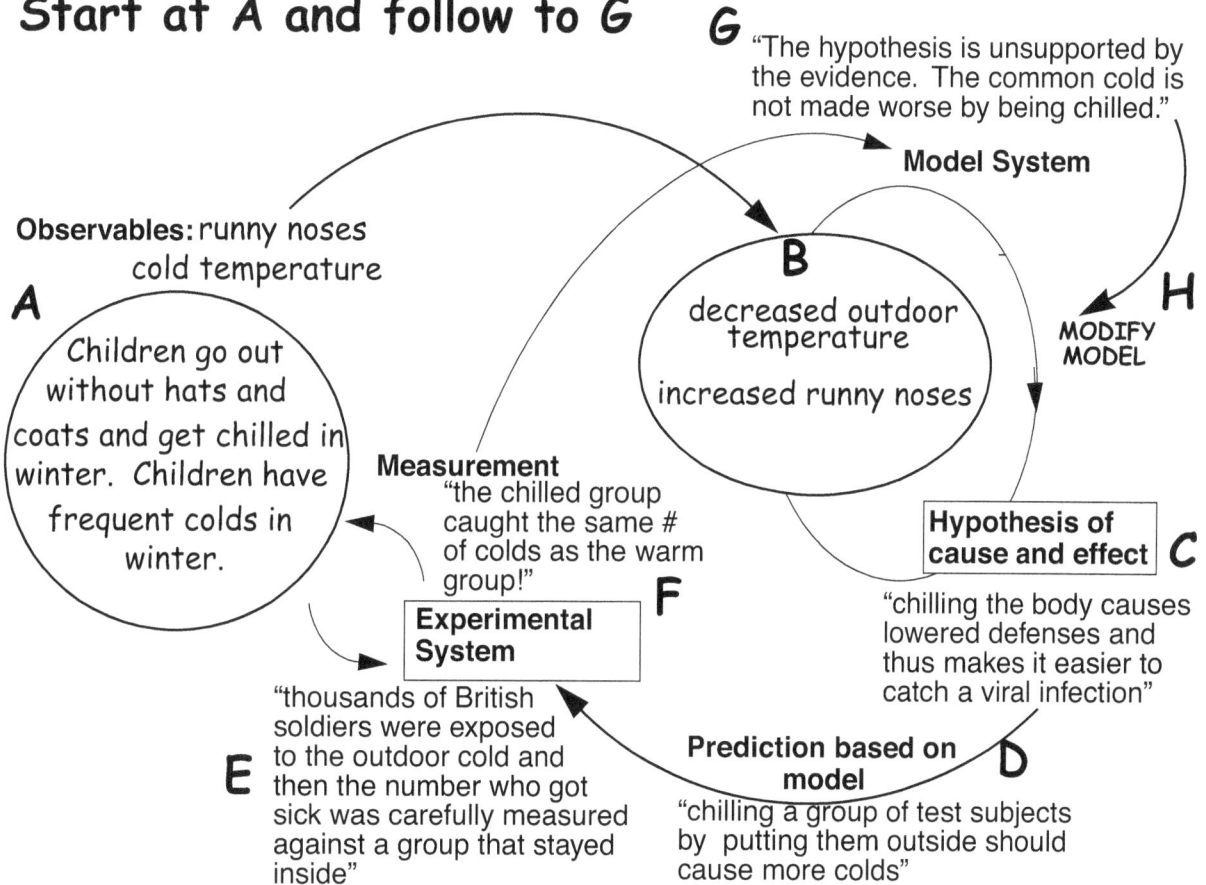

**G** "The hypothesis is unsupported by the evidence. The common cold is not made worse by being chilled."

**Model System**

**Observables:** runny noses
cold temperature

**A**

Children go out without hats and coats and get chilled in winter. Children have frequent colds in winter.

**B**

decreased outdoor temperature

increased runny noses

**H**

MODIFY MODEL

**Measurement**
"the chilled group caught the same # of colds as the warm group!"

**F**

**Experimental System**

"thousands of British soldiers were exposed to the outdoor cold and then the number who got sick was carefully measured against a group that stayed inside"

**E**

**Hypothesis of cause and effect** **C**

"chilling the body causes lowered defenses and thus makes it easier to catch a viral infection"

**Prediction based on model** **D**

"chilling a group of test subjects by putting them outside should cause more colds"

Scientific investigation has shown that it is not exposure to cold temperatures that causes colds. Further studies have shown that the common cold virus is spread when a "sneezed on surface" inside the house is touched. When the contaminated hand touches the mouth, nose, or rubs the eyes, the cold virus is transmitted. This leads to the hypothesis that hand washing should prevent colds. Studies have shown that hand washing is very effective in preventing the common cold.

## The Scientific Method and the Progression of Inquiry

Let's revisit our sequential modeling diagram and incorporate an experimental model as a requirement to verify and validate both observations and the explanations that the brain uses. Validation and verification steps that are derived from magical, "intrinsic/feeling-driven" or a philosophical mindset are not acceptable in modern scientific inquiry. We will reserve the phrase, "Progression of Inquiry" to be the inquiry process that specifically has an experimental model employed in a skeptical fashion as its validation process.

**FIGURE 11.** The progression of inquiry incorporates critical mapping (a verification process) at each step.

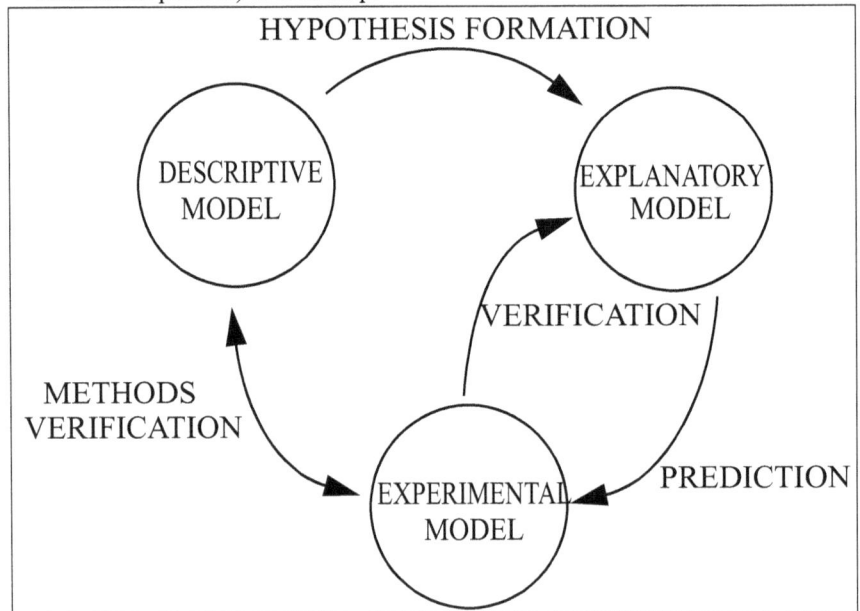

# CHAPTER 6

# The Human Brain: From Interaction to Cognition

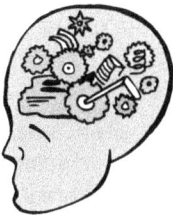

For all of its complexity of operation, the human brain is straight-forward in its outcomes. It searches and finds the *patterns* in what it senses. It extracts the features of patterns from the data. It assembles those features into a *model* of the perceived object. The Cybernetic Cycle captures the steps in the hierarchy from stimulus through the behaviors that the brain performs.

## Pattern Analysis and Model Making

All brain operations involve pattern analysis and model making. The brain has evolved to do a specific job of gathering and processing stimuli as an information stream from the outside natural world, and from the inside world of the body and mind. It uses the information to control the inner world of the body, and to relate the inner world to the outer world.

The brain gathers information from the outer world through a variety of physical senses (stimuli):

- sight (visible light)
- smell / taste (chemical sensing)
- touch (physical motion and weight)
- sound (vibration of air).

## Data Creation

The **interactions** with the world and the **stimulus** at the receptors (or sensors) create **data** (FIGURE 12).

**FIGURE 12.** Pathways for data flow into the brain

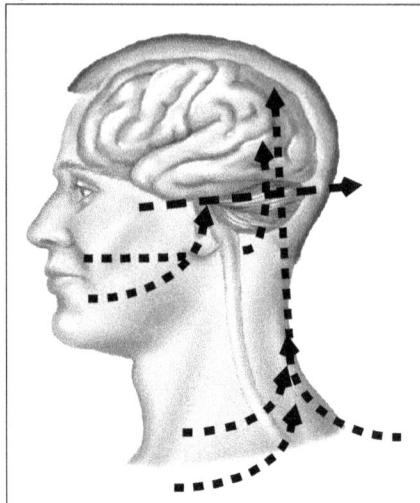

Let's consider an example - your dog, Rover. Rover can be seen, smelled, touched and heard. Since Rover is your dog, you carry other information about Rover, for example: his name, memories of the experiences that you have had with Rover, and so forth. The brain is organized into a series of nerve centers that act as pattern extractors and analyzers of information. Data is sent to these centers as a refined stream of **information**. Sensory experiences from the outside, natural world are represented in these centers.

**FIGURE 13.** Each of the primary senses has its own center or region.

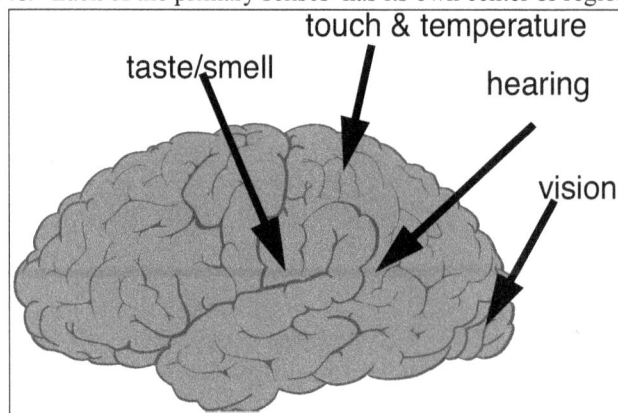

touch & temperature

taste/smell

hearing

vision

## Data Integration

There are also nerve centers in the brain that make memories, retrieve memories, and store memory information, such as the color, smell, shape, texture, etc. of Rover. When you think about, see, smell, or hear the name of "Rover", there are centers that associate all of those properties of Rover together! These are the cortical association areas. The brain is a very busy place!

**FIGURE 14.** Cortical association areas in the brain integrate information

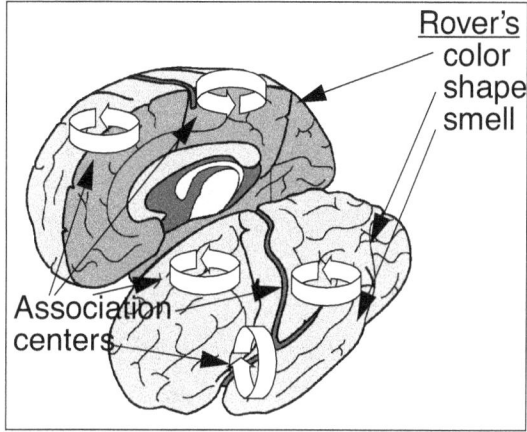

How does your brain analyze the information that it receives when you <u>see</u> Rover? For example, when your eyes focus the image of Rover, what does the eye tell the brain.

**FIGURE 15.** What happens to image information in the brain?

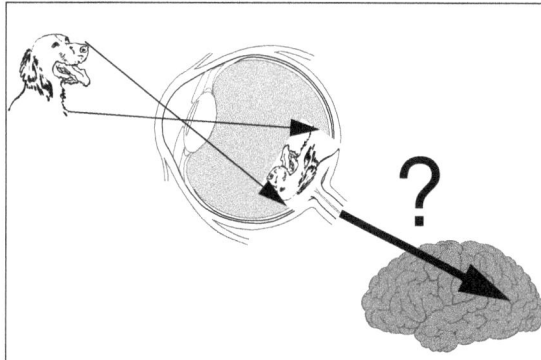

If the brain were like a digital computer or television, the image of Rover would be scanned and read point by point from the eye into the

visual center of the brain. Digital computers perform such point-to-point mappings in order to represent the image of Rover.

**FIGURE 16.** Information is digitally represented as a list in a point-to-point mapping (A) and as a list of digital numbers (B).

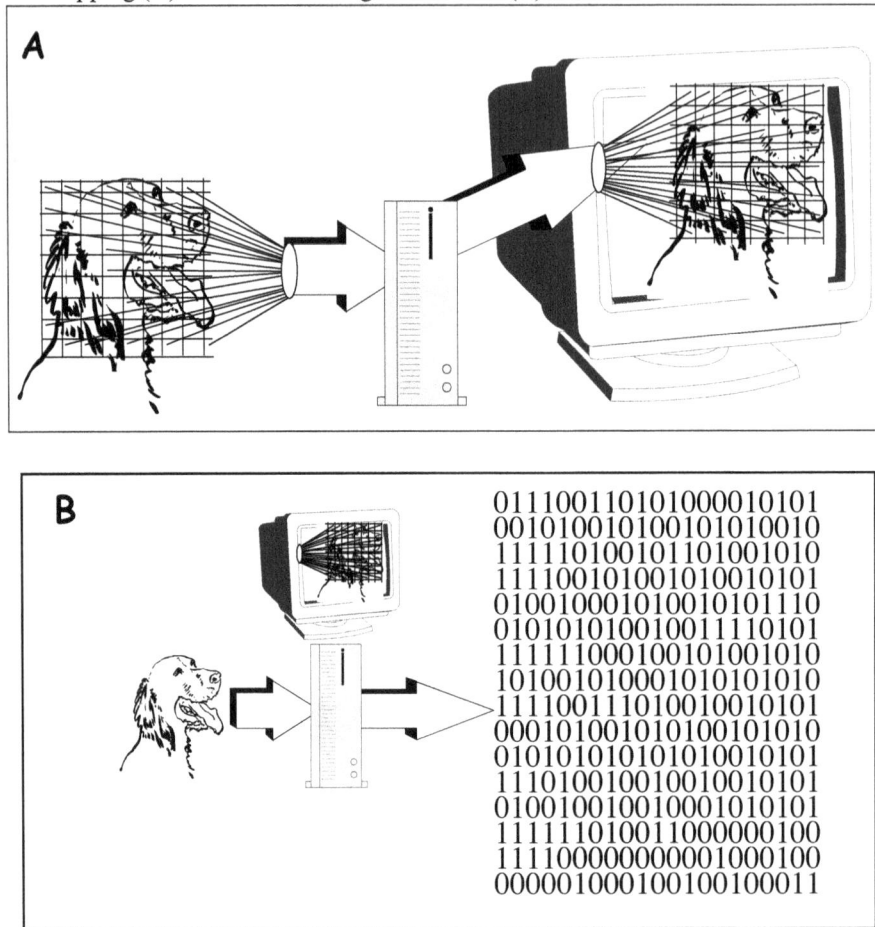

Rover becomes a long list of picture elements. The model of Rover in a digital computer is a list of numbers. In a digital machine, Rover is defined by this list. The list is processed by arithmetic (adding and subtracting). A change to any number on the list makes Rover a new object. Anyone who has ever made a simple typing error and gotten an "invalid file name" message from a computer has experienced this type of processing.

## Perception

**Perception** of sensory information in centers of the brain maps the elements of the external world into the internal model of the world. However, the brain does not work like a digital computer. Instead of scanning and processing information point-by point, the elements of an image are extracted as lines, contrasts of light and dark, curves, and even movements of a spot against a stable background. Perception is the detection by the brain of the system components of an object from the information stream it receives. In other words, the brain extracts the information from an image in terms of the *features* from which it is formed. These features are comprised of the *elements* (lines, curves, contours, etc.), the *rules* arranging the elements, and the *background space* into which the elements are arranged. These features are what we will call the **pattern of the image**.

**FIGURE 17.** The nervous system extracts features (A) and reassembles them sequentially in the brain. (B)

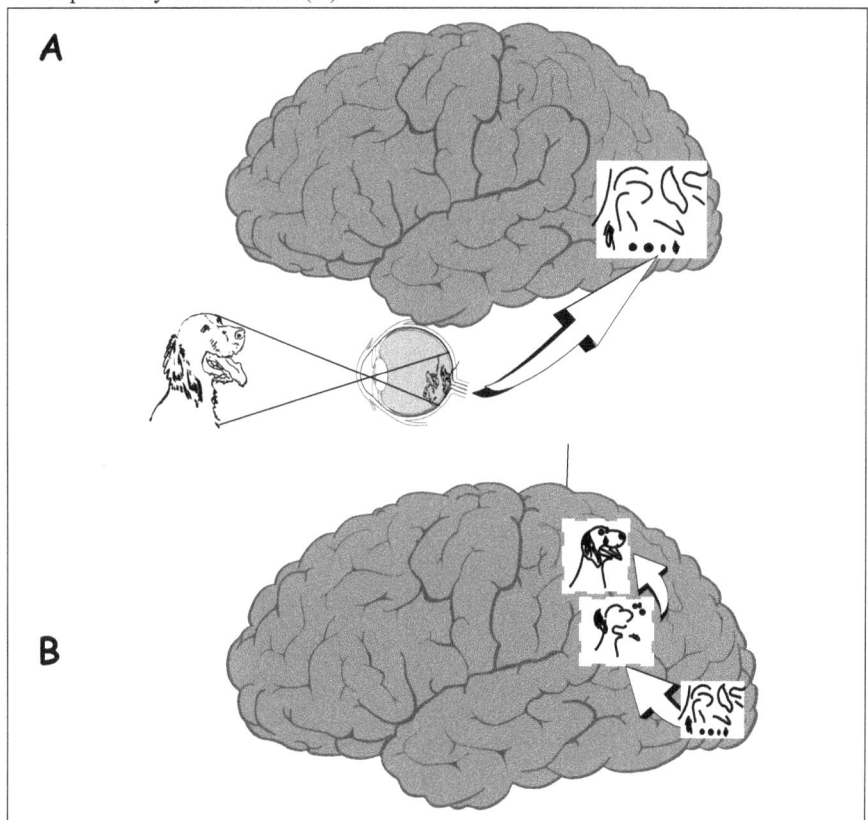

The brain constructs a model of Rover by reassembling the features of the information into a set of interrelated parts. This type of reconstruction allows Rover to be instantly treated as part of a group of Rovers. Changing a feature in a small way, such as Rover's ears being up rather than down, will not change the general ability to recognize Rover. Processing information in this way allows the human brain to read the misspelled file name and to still know which file to get!

As you can see, this way of processing information is very different than that of a digital computer. A digital computer records, stores, and processes every bit of information by counting and by rote memory recall of the arrangement of each piece of information. In contrast to the digital machine, the brain is a pattern extraction machine that evaluates everything by pattern analysis.

## Cognition

**Cognition** is defined as the point when a model can be brought to mind. Typically cognition is manifest when something is named or located in space. All brain-based operations work as follows:

FIGURE 18. The pattern analysis - modeling cycle. of cognition.

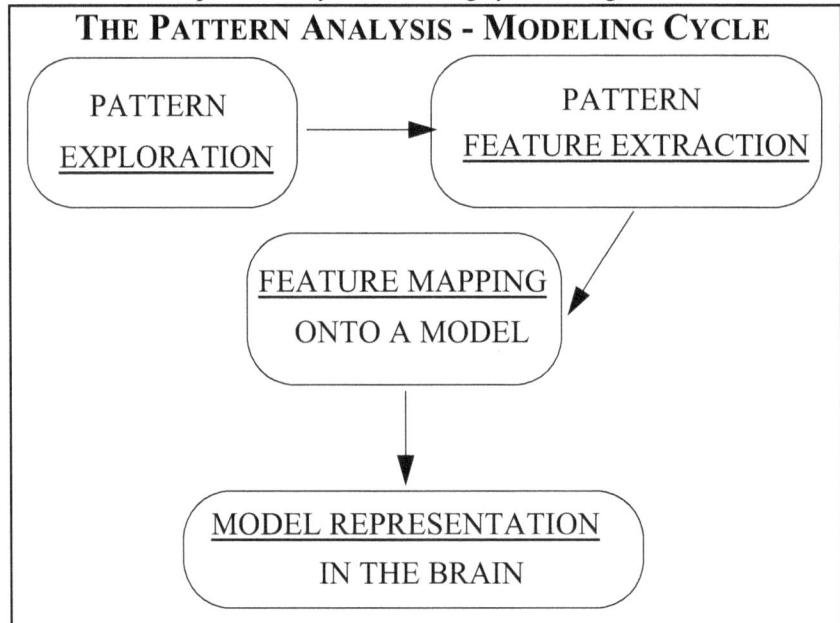

**THE PATTERN ANALYSIS - MODELING CYCLE**

PATTERN EXPLORATION → PATTERN FEATURE EXTRACTION

FEATURE MAPPING ONTO A MODEL

MODEL REPRESENTATION IN THE BRAIN

The brain therefore "knows" something by building a model of the object, person, action, or idea from its basic forms or pattern elements. Different parts of the brain are actively involved in the construction of the model and the knowledge that it represents.

This fundamental brain operation - the **pattern analysis-modeling cycle** - is repeated at many levels and operates in parallel modes in the brain. The power of the brain lies in the operation of these many different parallel modes and multiple hierarchical levels.

Continuing our consideration of Rover, we will explore how the use of parallel modes allows a complete model of Rover to be built.

## Knowledge

**Knowledge** of something is the integration of the cognitive information about an entity, including its associations with other entities often over time.

Consider Rover again. The visual features of Rover that were abstracted from your eye now reside as pattern elements, rules, and space in the visual centers of your brain (A).

The pattern of physical properties (touch, temperature, weight), chemical (odor), and sound properties (panting, barking, whining) also reside in their respective brain centers (B).

**FIGURE 19.** The brain integrates a range of information.

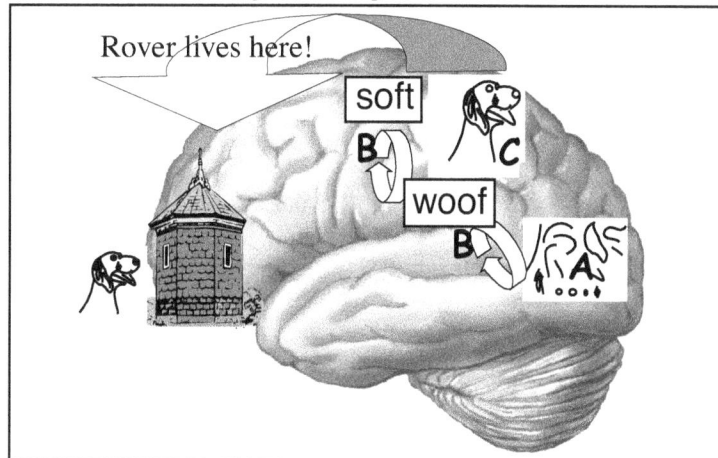

The primary sensory centers are connected together through association centers that take the extracted features of Rover and map them into a complete construction of Rover or "Rover's" model (C). However, this is not the whole picture.

While Rover himself is being constructed, the brain is also busy building a three-dimensional model of the space in which you and Rover live. Rover is then placed in the space that the 3-D space centers have built. <u>Rover lives in a model space that you have built!</u>

The models can be manipulated, sorted, categorized and compared in connected parallel regions with other information - also represented as models of pattern extracted features. There is an advantage to the information processing strategy of the brain - the simple elements of "DOG" and the arrangement of those elements gives rise to a huge group of "DOGS" of which Rover is just one.

**FIGURE 20.** The basic elements and rules that compose Rover are arranged just slightly differently. Each of these animals is clearly a dog yet they are not all Rover.

On the bottom row, every dog is clearly Rover, yet a point by point comparison of each dog (as a digital computer would perform) would not recognize the models as representing the same animal!

The pattern feature analysis mechanism of the brain allows broad categories or groups of objects to be easily formed and sorted. The ability to classify and sort is a very important power that comes from the organization of the brain. Pattern analysis in the brain provides a pattern of relationships between the connected association areas. This "pattern of patterns" provides input to a higher brain-center feature analyzer. This analyzer is then able to conclude that the object seen is a dog (not an insect or lion or cat or bird, etc.).

**FIGURE 21.** The relationships between various properties lead to the identification of DOG.

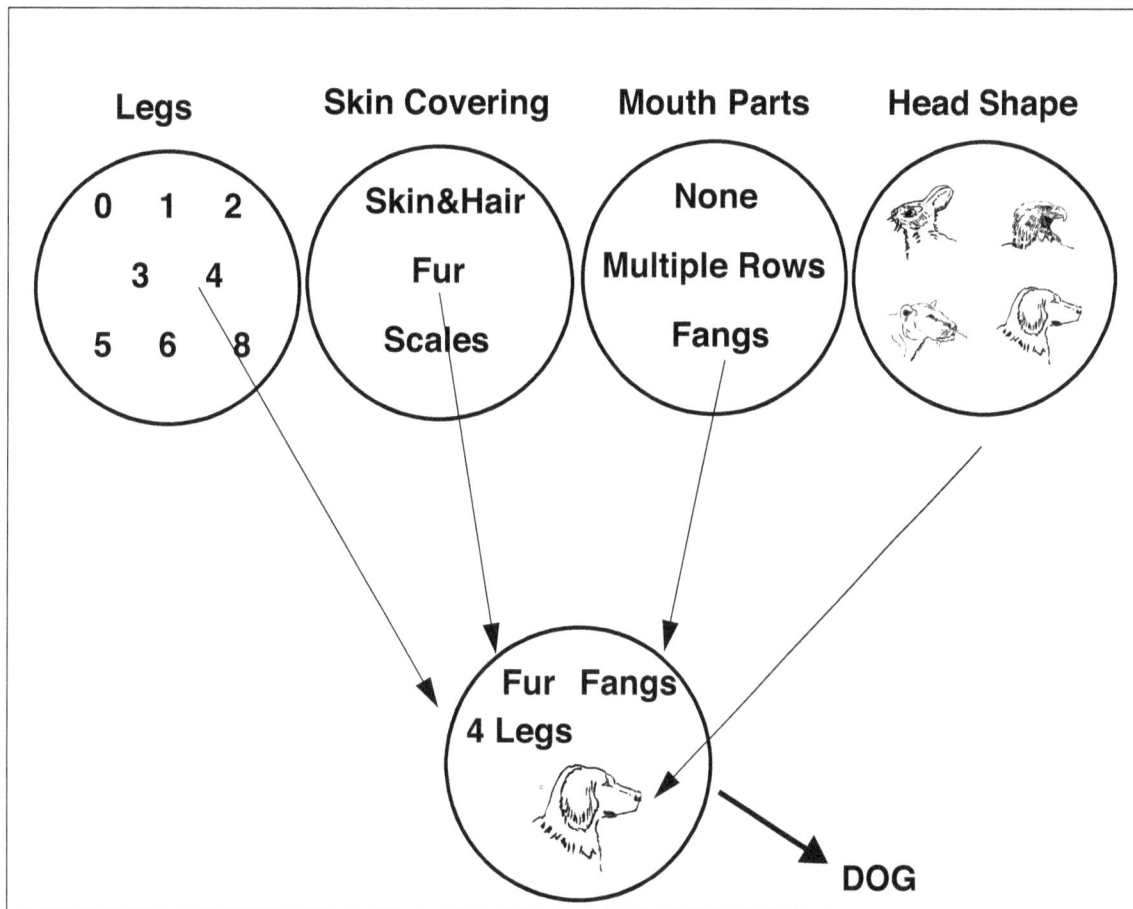

When matches are found between patterns of Rover, still another nerve center will tag the object - -DOG - with the symbolic name: "DOG-Rover".

If an association center presents the model of "DOG - Rover" to the speech centers, the words and connected ideas about "Rover" can be expressed: "That is Rover, my dog".

**FIGURE 22.** Words can be associated with the internal model of a cognate generating names with semantic meaning.

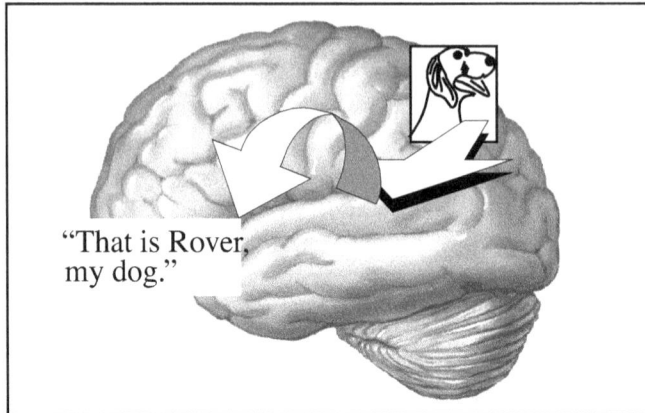

The brain uses pattern features to construct models of objects from observations of the outside world and also from memory. It also can build models of objects in the imagination. Both real and imaginary models are constructed by the brain and interact in a background space created in the mind. Once the brain makes a model, it is usually regarded as complete, (real), even when it comes from the imagination or from a hallucination!

**FIGURE 23.** Models in the brain are treated as real by the mind.

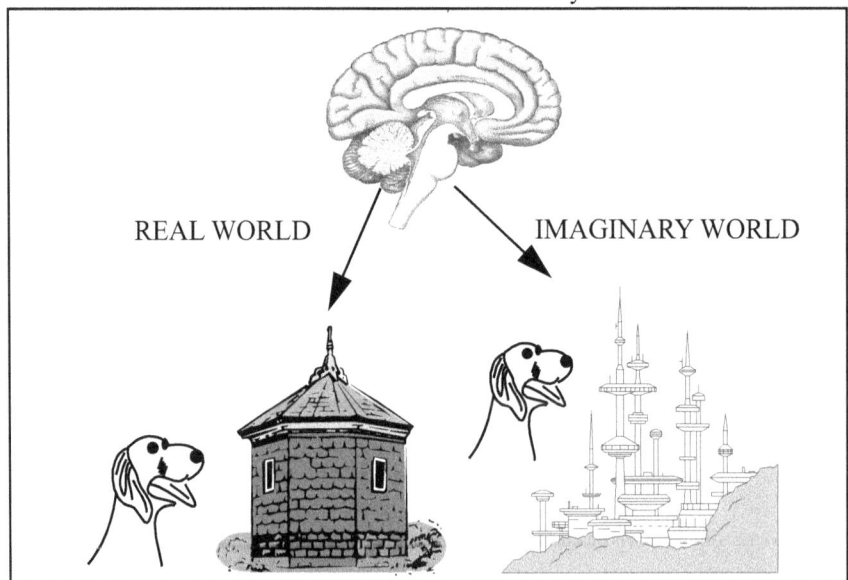

## Understanding

**Understanding** is the application of knowledge. Understanding means being able to apply the knowledge contained in one internal model to another internal model. The interaction may be intended to solve a problem in the real world, make a decision or judgement, or engage or create a new entity.

**Habit** and belief are the patterned response to familiar interactions. At each step the initial stimulus-generated data stream has become less discrete but more knowledge laden. After a series of knowledge associations has been applied to similar (or the same) problem, the brain simplifies the complex process of solving the problem through creative understanding. It abstracts the important elements of the solutions and makes them habit. 'Habit', 'muscle memory', 'enculturated rites and rituals' are all part of the output of a specific part of the brain (see Appendix A). Belief and habit are the organized almost automatic responses to patterns of stimulation that the brain recognizes through the process of the cybernetic sequence.

# CHAPTER 7

# From Brain to Mind: Constructing Knowledge and Understanding in the Brain

## Cause and Effect Patterns

The human brain is unique in its quest for a pattern of cause and effect. Cause and effect patterns are the basis of artistic, religious, and scientific approaches to the world. They require a "mode of explanation". They answer and explore the questions of:

- "Why?" and "How?"

- "What if?"

- "What is my relationship to the world?"

- "What is my purpose?"

In cause and effect patterns, the system features are:

- emergent properties - characteristics of the overall system.

- elements - two or more components of the system, linked by a function or rule.

- a rule - of the cause and effect usually reflected in a mode of explanation.

- a context (background space) - in which events occur.

In the brain, cause and effect is computed in the same fashion as a pattern of lines and contrasts. The difference is that the elements of the pattern are related by <u>inferred explanations</u> rather than to observed relationships. The brain builds a model to explain cause and effect using the same neural mechanism that is used to analyze the visual patterns of "Rover".

Since the brain works by pattern feature-extraction and model-making, several important behaviors result. The brain tries to resolve ambiguity when constructing a model. The brain completes a model created from a pattern even without adequate information. After creating a pattern from inadequate information, the brain may discard the original sensory data and then indiscriminately attribute "reality" to an incorrect model.

## Ambiguous Patterns

The brain avoids ambiguity whenever possible. Sometimes it is not easy to avoid ambiguity. Here are several well known examples of figures that are cognitively unstable - it is hard to determine which object or what three dimensional orientation is being shown.

**FIGURE 24.** Examine these ambiguous figures.

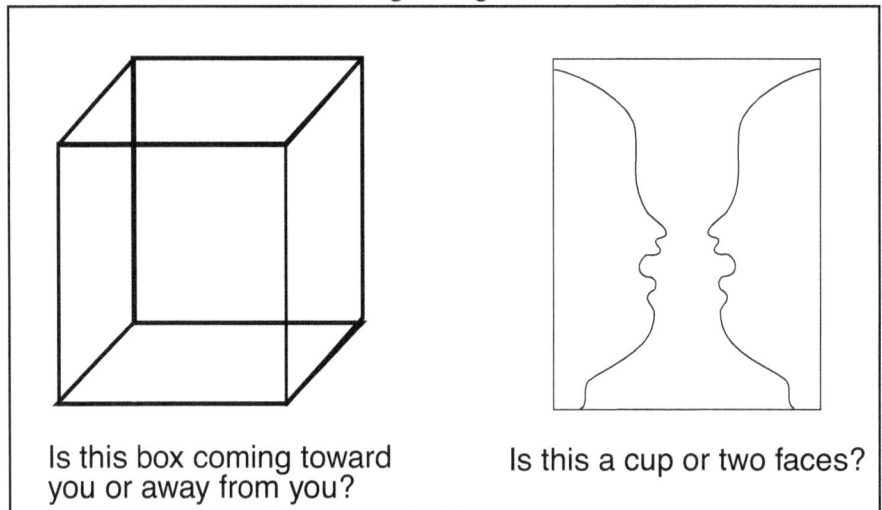

Is this box coming toward you or away from you?

Is this a cup or two faces?

These examples of ambiguous visual figures are often unsettling to viewers. These are illusions of cognition. The brain can not lock onto the features needed to construct a non-ambiguous figure. If possible, the brain will attempt to stabilize and hold onto one of the images. These particular test figures are designed to make such stabilization very difficult.

## The Knowledge Surface

A great deal can be learned about the brain works from these figures. A model of how the brain operates will explain these peculiar effects. This complex topic is more easily understood visually by using this diagram. Let's do what the human brain needs to do...make a model.

The type of surface shown in FIGURE 25 represents the behavior of the brain. It is an example of a knowledge construction surface. We will call it the knowledge surface. A surface with this shape can explain the effect that you experienced in the earlier pictures.

**FIGURE 25.** This cusp surface is a mathematical model of brain behavior.

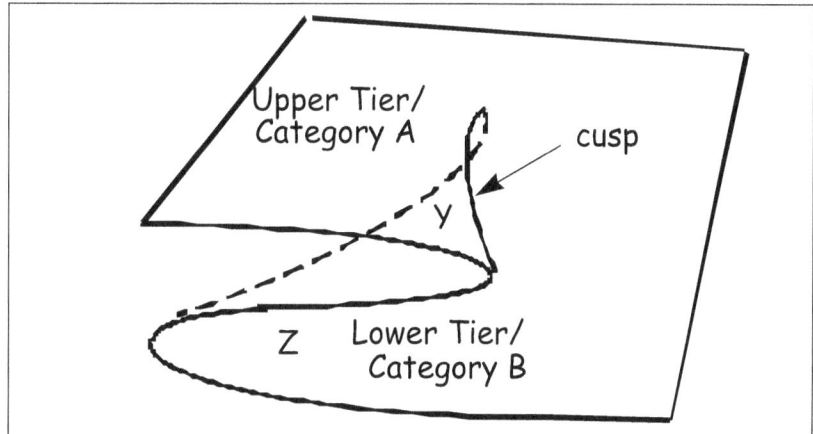

This knowledge surface represents the way the connections in the brain work together to classify the information. Notice that there are three main features.

· Two tiers that each represent a category.

　· A "cusp" that projects one surface over (Y) and the other surface under (Z) each other.

　· The brain always prefers that an object is assigned to one tier or the other.

## Ambiguity

Pattern features are assigned to a knowledge surface. In the case of an ambiguous figure, the brain builds a model with conflicting information and becomes confused.

**FIGURE 26.**  Look at the possibilities available to the brain in deciding whether it is seeing a cup or two faces.

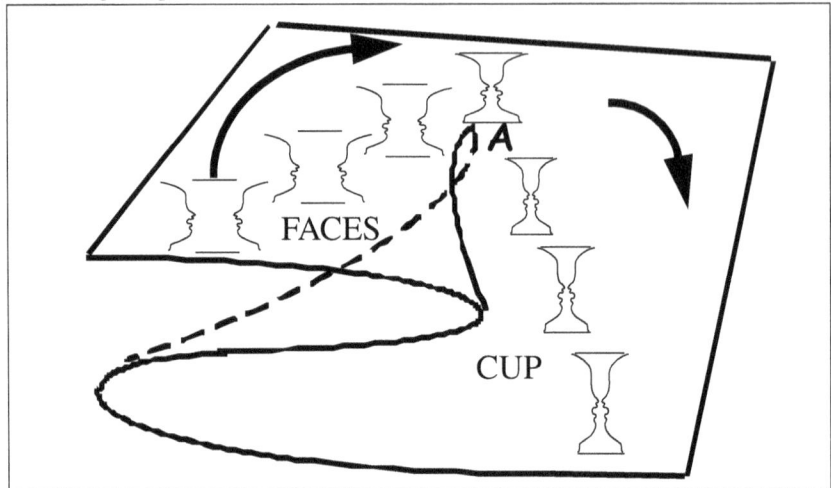

The shape of the surface allows many cups or pairs of faces to be recognized as stable. A large variety of different pattern feature arrangements will lead to the "knowledge" that two faces are seen. These possibilities for faces are represented by the upper ledge, on the left side. Alternatively, the possible arrangements of the pattern features that form the cup are represented on the surface on the lower right.

A distinctive characteristic of the knowledge surface is the "cusp" in the center of the surface. The pattern feature arrangement leading to the ambiguous figure falls on the top of the cusp at point A. The region at point A has the behavior of appearing alternatively as faces or as a cup.  The easy flip-flopping between the two modes - faces and cup - is an unavoidable consequence of the brain's structure.

What does this "knowledge surface" tell us about the behaviors of the brain and mind? Examination of the surface shows several striking qualities. Let's call on Rover to help  again. (FIGURE 27).

The circuitry of the brain  is designed to  group or classify a set of features into a  category. As features (A) are mapped onto a knowledge surface (B), categories are established. These general categories are represented on the two planes of the pictured  surface - "DOG" and "BEAR" (a model of the complete brain would have many more category spaces).  There is a fairly broad range of pattern arrangements that fit into a group of similar DOGS or BEARS. This flexibility gives the upper and lower tiers their shape and size.

**FIGURE 27.** Differences between dogs and bears.

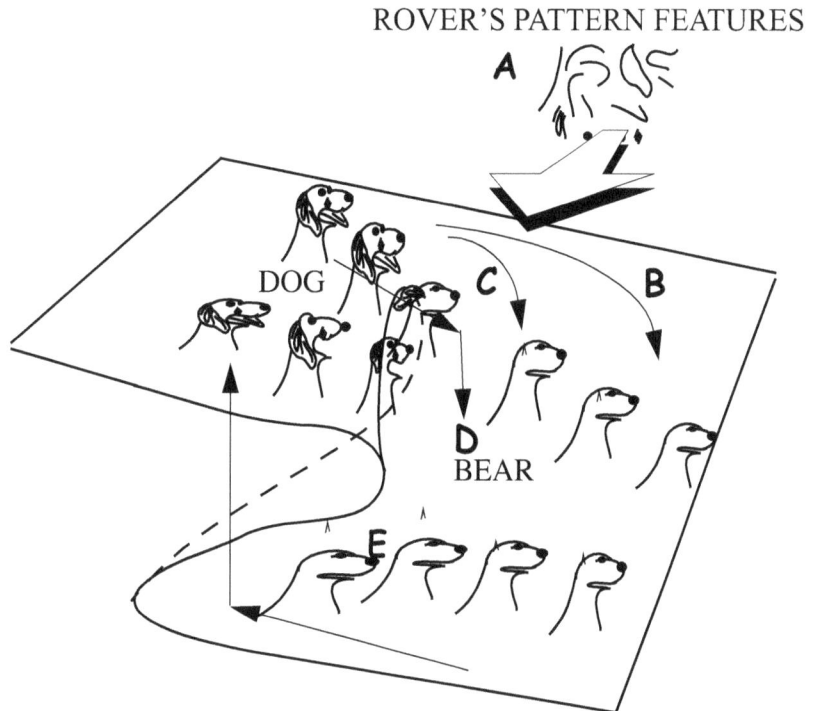

ROVER'S PATTERN FEATURES

In order to change from one category to another, a path must be found that transforms the pattern features from one group to the other. A variety of smooth paths (point C) to make this transformation can exist by going around the top of the cusp.

The overhang and the underjut of the cusp surface however give rise to a very important behavior. It is not easy to change from one group to another near the cusp (point D). Going from the upper ledge to the lower tier is resisted because the upper ledge juts out over the lower tier. Then a small change in ear shape causes the cusp to be crossed and the group identity (DOG--->BEAR) changes suddenly, in a flash - this is the "Ah Ha" or sudden insight that is familiar to all people.

Likewise the underjut resists change from lower tier state to upper surface (point E - here the head shape flattens a great deal before the bear changes into a dog). These properties of the cusp are a reflection of the nervous system's tendency to "lock" onto a particular state or idea.

Grouping and sorting into categories is a natural strength of the brain. The knowledge surface has different stable regions that represent the set of objects that share the features of a group (e.g., tiers A and B).

## Bias and Prejudice

Another important behavior of the knowledge surface can be experienced.

Look at the following picture. Start on the left and cover the images to the right of your view. Uncovering each image one at a time, how far can you go seeing a man's face?

**FIGURE 28.** Read this figure from Left to right.

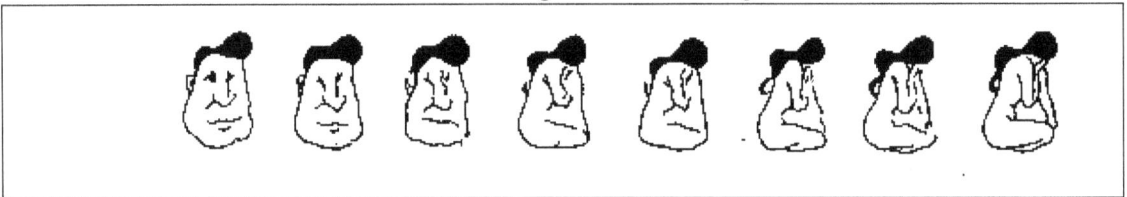

Now start on the right. Cover the images to the left of your view. Uncovering each image one at a time, how far can you go seeing a woman's body?

**FIGURE 29.** Read this figure from right to left.

Notice that the point where the change from one category to the other occurs, depends on which direction you started from. You are experiencing the resistance of the cusp region to changing the "state of being" of the knowledge surface.

The example demonstrates first-hand how the mind locks onto and holds a series of objects in a group until suddenly the group features no longer can place the object in that category. How this experience can be modeled onto the knowledge surface is shown in the figure below.

**FIGURE 30.** Mapping of the experience of reading man/woman series.

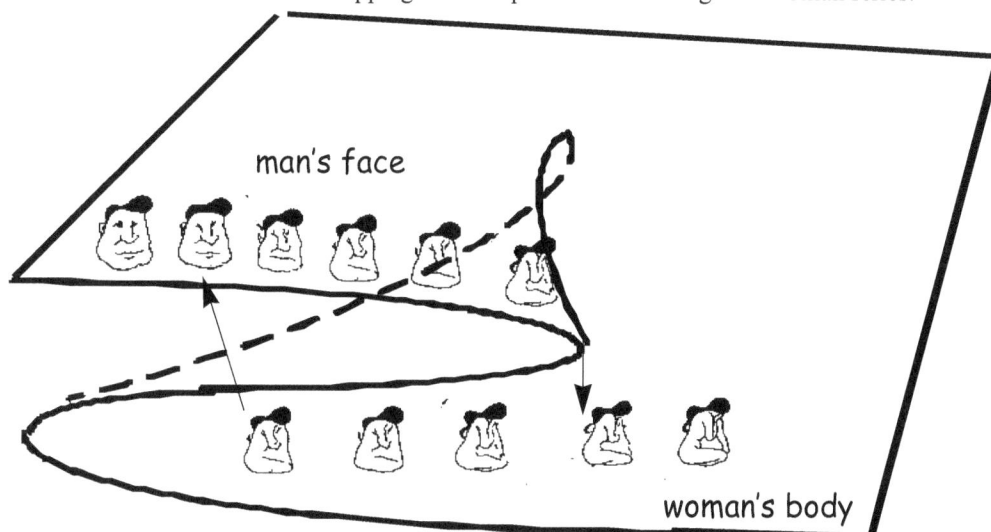

This exercise demonstrates how bias emerges in perception and cognition. The bias is due to the cusp property in "knowledge surfaces". The cusp is a natural consequence of the structure and organization of the brain and derives from the way the brain is connected together. The cusp's properties explain how *prior knowledge and expectation* influence these systems. Since the cusp in the knowledge surface is intrinsic to the organization of the brain, it explains the natural tendency of mind toward bias and prejudice as important factors in learning and behavior.

This example demonstrates that while the shape and properties of the knowledge surfaces can be modified by learning, the general properties of bias and tenacious adherence to the status quo are intrinsic. They can not be eliminated. They are structural in the biophysics of the brain. However, through the scientific and critical thinking process, these tendencies can be detected and mitigated.

Ambiguous figures are usually resolved as one or the other figure, even when unstable (moving back and forth between the two images). This bimodal property of the nervous system is a reflection of its need to "lock" onto a particular state.

The circuitry of the brain prefers stability over accuracy. In the case of ambiguous information, it will do this by either ignoring some information or by increasing bias to lock onto a stable model. Ignoring information reduces the fineness of the data presented to the brain. The loss of detail and information reduces the ambiguity. The brain will sacrifice detail to find a stable state if necessary.

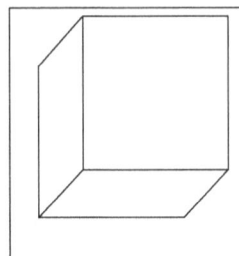

"Expectation" conditions the brain to enter one state or the other. This type of conditioning is how a preconceived notion or perception can strongly influence what becomes known in complicated situations.

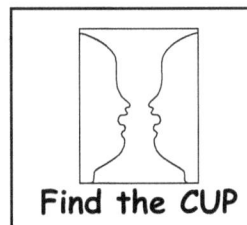

**Find the CUP**

## Modeling Learning

What about learning in our model? So far we have looked at a single snapshot of the brain. What makes the brain so important is that the brain is a learning machine. This allows experience to be treated like a movie, not a single picture. The brain is capable of altering its behavior and analysis of information based on previous and new events.

Throughout nature, all animal brains:

- make observations using pattern extraction

- store sensory experiences as pattern features in memory

- learn by associating the patterns of a stimulus with the observed pattern of its response

All animals, including humans, will show altered behavioral patterns as evidence of this type of direct stimulus-response conditioning.

Much has been learned about the complex process of learning in the brain in recent years. It is clear that the process of learning changes the strength and types of connections between the nerve cells in the brain. It is the organization and interaction of the nerve cells that gives the knowledge surface its characteristic shape. While the fundamental form of the knowledge surface is fixed by the structure of the brain (and its cells), learning alters the specific twists and turns of a knowledge surface as the number and type of connections themselves are altered.

How can we model learning?  Rover will help us, again. Learning occurs when events are connected more reliably one to another. Earlier we noted that the shape and size of the tiers reflected all of the various ways that an object could be stretched and twisted and still belong to the same group.  A very large number of ways has a large tier with each object connected to the others.  A small number of ways of being represented will have a small-sized tier.

Consider what happens as we learn what a "DOG" is.  If Rover is the initial experience that defines the category of "DOG", his features will be mapped onto a knowledge surface with a fairly small tier representing DOG.

**FIGURE 31.**   The knowledge surface expands with experience.

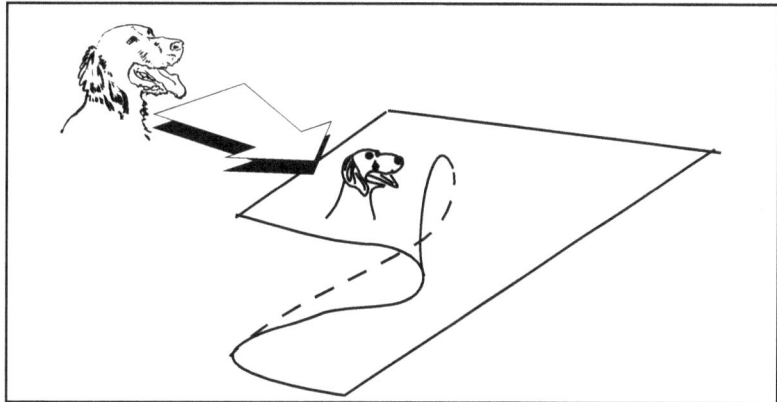

Because the knowledge surface has a certain inherent flexibility, even when it is small it can accommodate nearby-related objects that it will treat as Rover-like, i.e., DOGS.

FIGURE 32. Expaning knowledge surfaces find similarities in diversity.

This means that if we were to see a "Rover-like" animal, we would classify it as a "DOG"

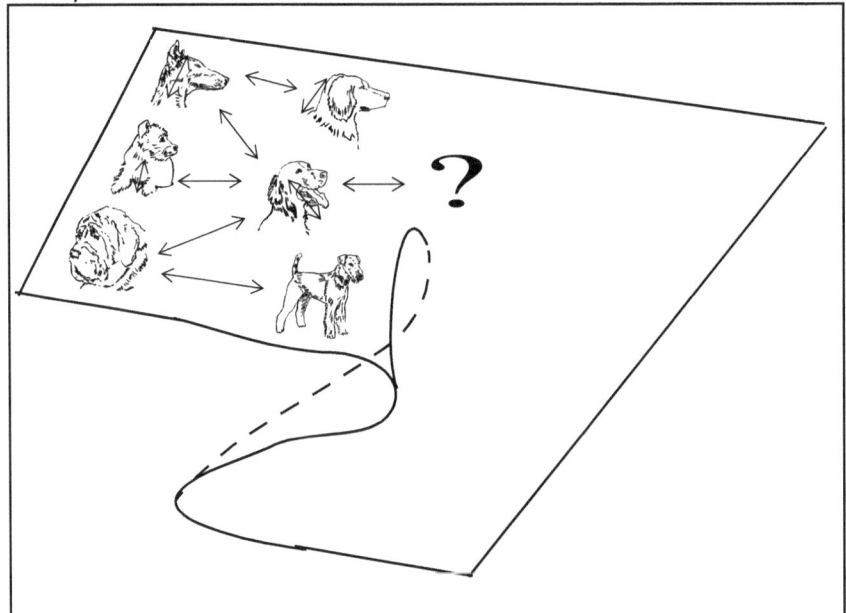

With a greater experience with "DOGS", the brain begins to modify the area of the DOG category surface. As the surface grows, the category of DOG becomes richer and more nuanced.

Learning is, therefore, the process of changing a category space (like DOG) on a knowledge surface. The changing size of the category region in a knowledge surface accounts for an important observation: that people learn more easily when a subject or its context is already known. A new general category space is established by first learning (mapping) a specific experience onto knowledge surface. Then the set of specific experiences is expanded to a more general case. This more general case is an abstraction. An abstraction represents a broader area of a knowledge surface.

## Changing Knowledge Surfaces

Knowledge surfaces change with learning. Where you begin on the knowledge surface has enormous impact on how the knowledge is constructed. Let us consider the case of how a teacher might judge the classroom actions of the "teacher's pet" versus the "class clown", as depicted in the series of knowledge surfaces in FIGURE 33 (figures 33a, 33b and 33c).

When an episode of talking in class is observed, whether the teacher assigns it to the "good" or "bad" behavior category depends on the degree of the behavior and the type of student doing it. Clearly a small amount of talking (A) by the "class clown" will be treated as bad behavior, while a much greater degree of talking (B) by a "teacher's pet" will be "forgiven" as OK.

Over time, if the "teacher's pet" begins to take advantage of this particular instructor, the category surface may change as the teacher learns this. Now behavior that previously would have been forgiven (B) will be re-categorized as "bad".

**FIGURE 33.** As the shape of the knowledge surface changes, the cusp portion also changes, as seen in FIGURES 33A, B, AND C.

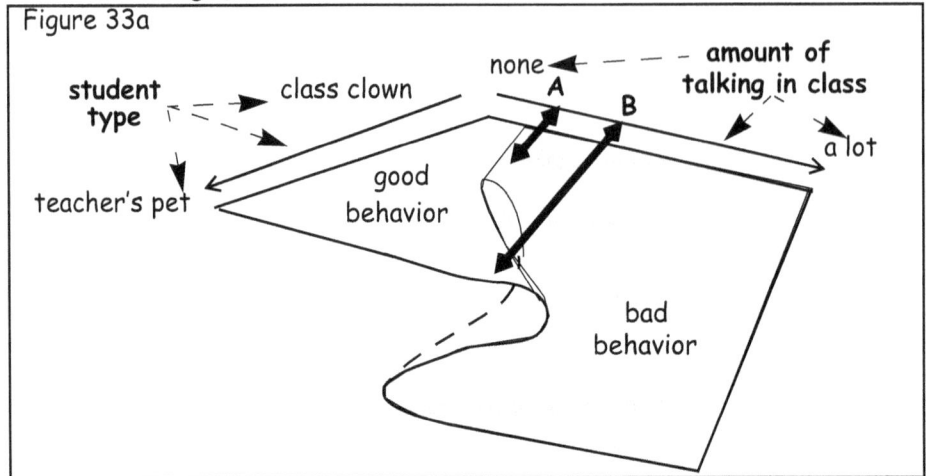

Figure 33a

The next figure (33b) shows this change as a dark grey zone. This zone appears as the teacher reconstructs the model of the students in the class. The cusp narrows and the overhang that earlier resulted in tolerance for the antics of the class clown no longer "forgives" the talking behavior at (B).

Note that the change in behavior at the "teacher's pet" end also results in even less tolerance for the "class clown's " behavior (A).

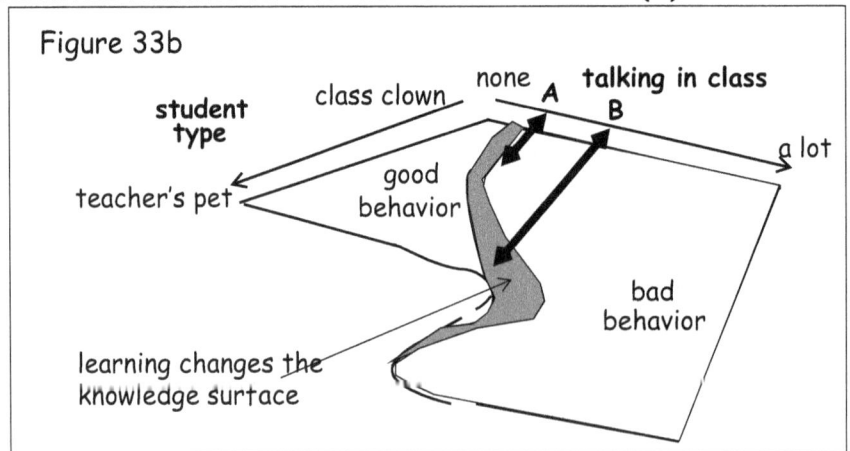

Figure 33b

Once the cusp is crossed, and bad behavior is expected, it will take a significant change in the classroom behavior of the students before the instructor will again view the "teacher's pet" behavior as good.

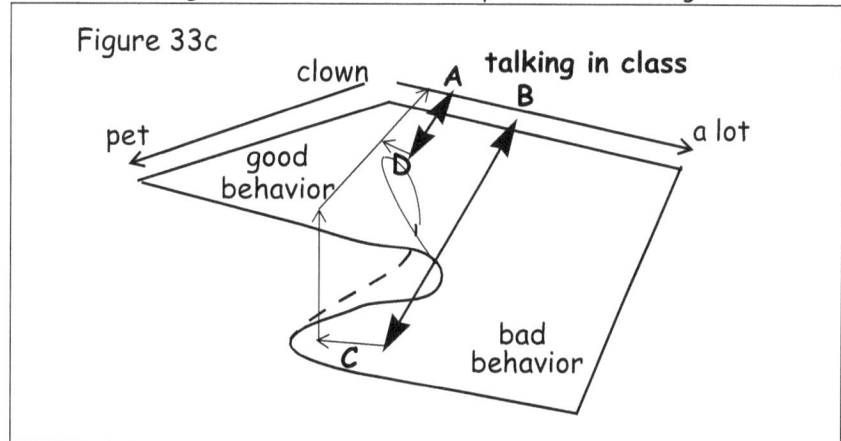

Figure 33c

In fact, it will take much more improved behavior on the part of the teacher's pet compared to the clown to get back into the instructor's good graces. This is represented by the long arrow showing improvement (represented by decreased talking) by the "fallen" teacher's pet (C) compared to the much shorter arrow (D) for the clown.

Because the cusp shape and depth determines the tendency of the brain to hold onto an idea or to be biased toward a particular viewpoint, learning can alter the ease with which a knowledge surface can change its world view or viewpoint. This is why the proper context makes learning easier. Alternatively, information that is learned incorrectly will be resistant to correction.

## Snap Judgements - the Aha Moment

We have already noted that the "Ah Ha!" experience is explained by the shape of the knowledge surface resulting in fuzzy information processing. This is a similar case to what happens when someone makes a "snap judgement".

Everyone has had the experience of making a quick judgement of someone or of a situation in a very fast brief encounter. These judgements are made on a very limited set of information - "first impressions". A judgement is simply a categorization onto a knowledge surface. Not only are such judgements (categorizations) made with very small, incomplete sets of information, once made, it is very difficult to change the judgement or category into which the person or event was originally placed. This is a very good description of the behavior expected from our model knowledge surface.

Where did such a tendency to make "snap judgements" originate? In the evolution of the brain, it was a frequently necessary to recognize a dangerous situation with limited information. In the case of a predator or a dangerous environment (like a cliff), collecting a complete set of sensory information is often possible only when the predator or cliff are within fatal striking distance. The brain had to evolve the capacity to "recognize" a dangerous situation before a complete set of pattern features was knowable in order to survive. The interplay of knowledge surfaces can account for this well known behavior.

At every stage of processing in the brain, the abstracted pattern is categorized onto a knowledge surface, thus building a model. As a model is being constructed, the process of categorization is proceeding in parallel. There is feedback between these two knowledge surfaces, one building the perception, the other proceeding with real-time categorization (FIGURE 34a).

**FIGURE 34.** The knowledge surface model can help us understand snap judgements.

Figure 34a

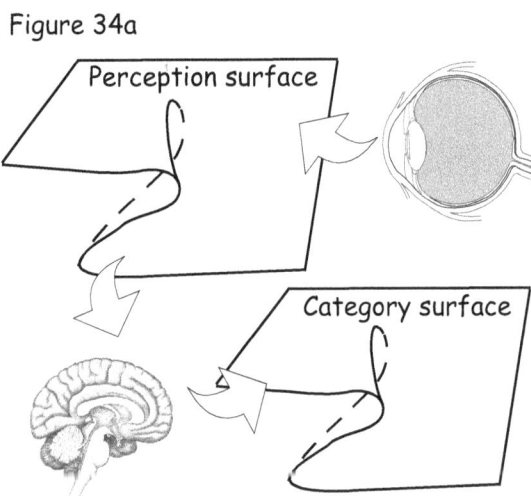

If the model being constructed is completely novel, i.e. being constructed for the first time, the categorization process occurs on a small and naive knowledge surface. The brain will not snap to a category assignment, because the naive surface does not have a strong category structure (FIGURE 34b).

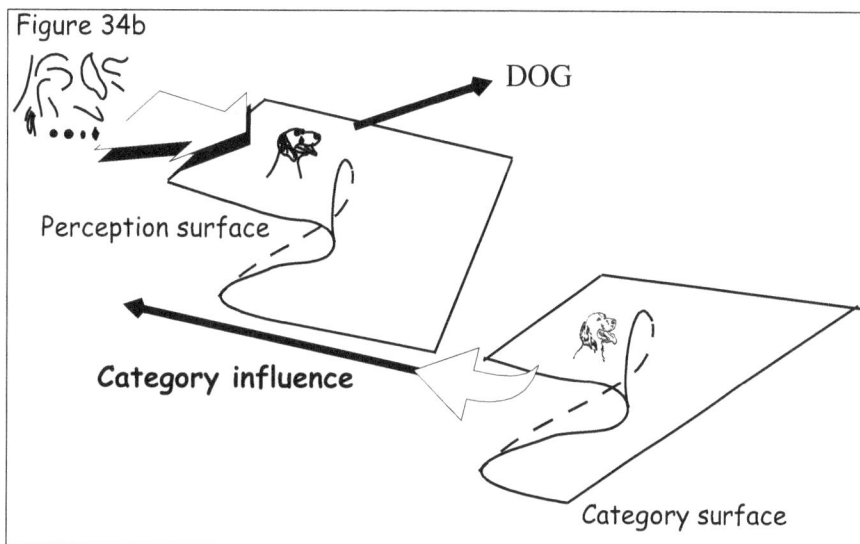

Figure 34b

However, when the model being built is being influenced by a knowledge surface with a high prior learning value, it will only take several features in the new observation model to confirm the presence of the expected feature (C), i.e. *My DOG*. In the case of the unexpected, if danger is expected (D), a shadow is likely to be quickly categorized as "dangerous". i.e. **BEAR!!**. An agitated dog is more likely to snap in fear when it expects that the approaching stimulus is something to be feared.

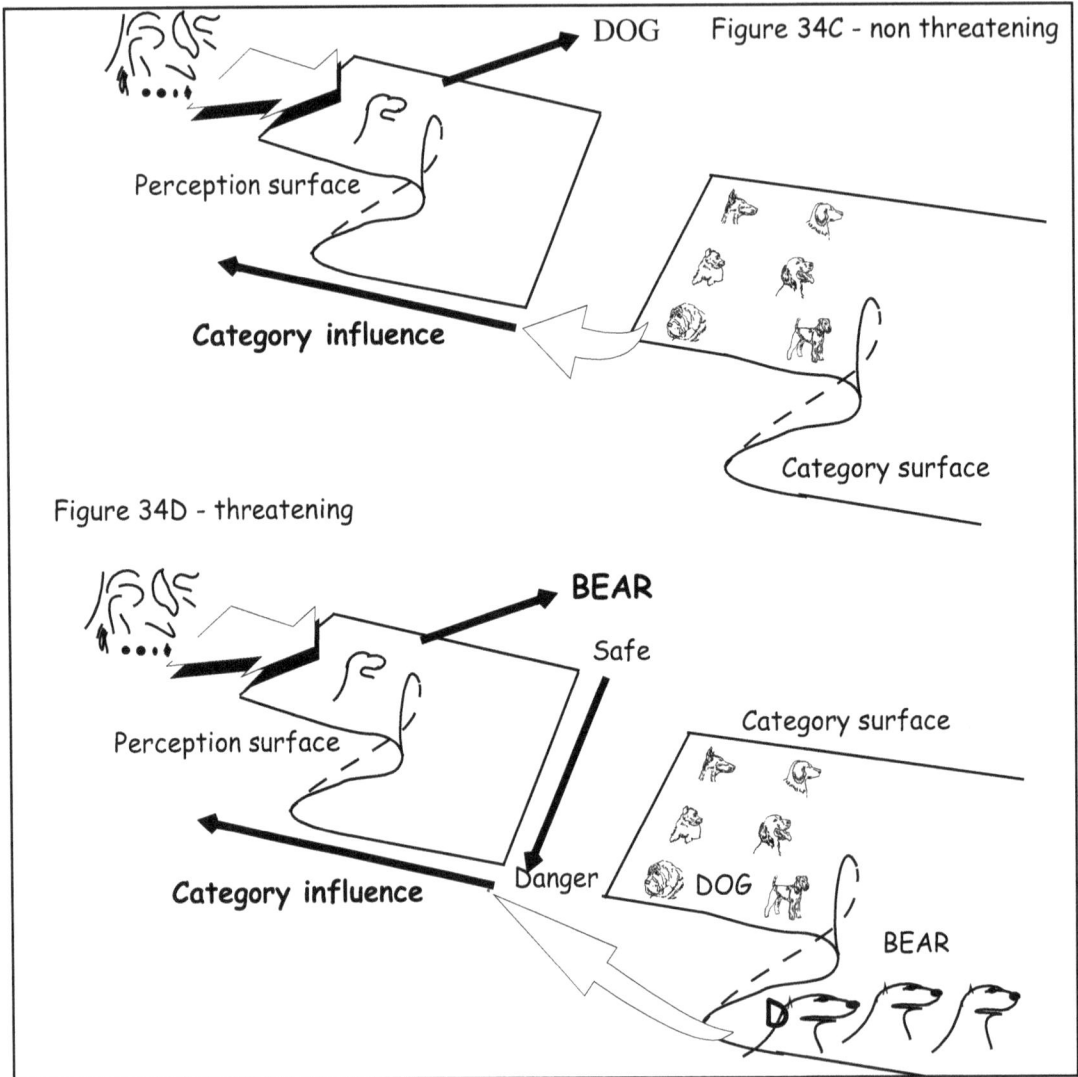

Figure 34C - non threatening

Figure 34D - threatening

A child whose imagination is filled with scary monsters, will "see" monsters under the shadowy bed because the limited sensory input is completed by the expectation of a "monstrous" categorization surface. This is why a blindfolded child in a carnival haunted house when told that a handful of grapes in floating in water are eyeballs is so easily scared.

The brain discards primary data and attributes truth to the model. The brain will complete a pattern or model even if insufficient data is available. If an ambiguous edge or shaded border can not be clearly detected, the brain will guess at a reasonable approximation and complete the pattern. Expectation and its accompanying bias will play a varying role as described above.

Once an incomplete pattern is assigned to a knowledge surface category, the assigned pattern gains all of the attributes of the category. Therefore, once a model is categorized, the representation is treated as true. This occurs regardless of the initial level of ambiguity of the data. Even if the primary data could falsify the model, if it is discarded, the model will be regarded as correct.

This is why a lesson or habit learned incorrectly is so hard to unlearn. The assignment of a pattern to a stable knowledge surface carries with it the resistance to change inherent in the shape of the cusp.

CHAPTER 8

# The Growing Brain: Cognitive Development and Implications for Learning

## Knowledge Processing

The centers of the brain that are required for model construction and mapping are physically separated. Communication between these centers is essential.

**FIGURE 35.** Knowledge processing regions in the brain.

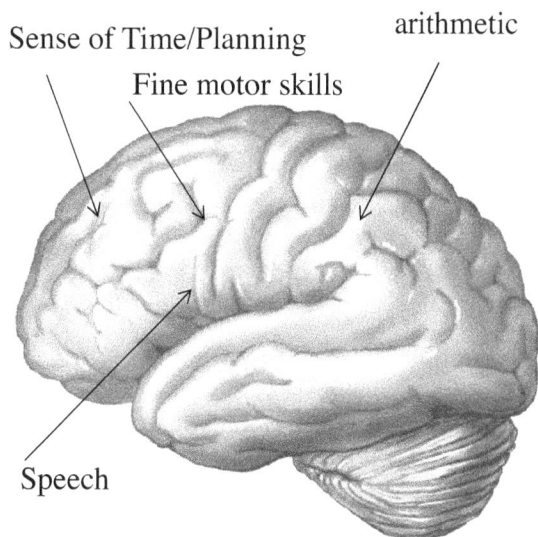

Sense of Time/Planning

Fine motor skills

arithmetic

Speech

The human brain is continuously maturing from infancy through early adulthood by physically connecting these various brain centers, which is complete by age 25. Therefore, not all brain centers are developed nor fully connected in young learners. For example:

·  The ability to understand and construct on very large scales (astronomical-millions and billions) or small scales (microscopic-millionths or billionths of an inch) require brain capacities that are not developed until the average early learner reaches adolescence. Children have a hard time understanding very large and very small magnitudes because they do not have the capacity to construct the proper models.

·  The ability to construct events into future time requires portions of the brain that are not fully connected to the rest of the brain until late adolescence. Children and young adolescents are not able to anticipate and effectively plan when future consequences must be taken into account, because they have not yet developed the required connections to construct a model of the future that is realistic and meaningful. This is why when a 14 year old boy promises to clean up his room "later", later never comes.

Both learning and biological maturation interact throughout the first several decades of human life to dynamically influence the final shape and flexibility of the knowledge surfaces of the mind. As maturation alters the overall structure of the brain (by its impact on connectedness), these maturing connections alter the shape of the knowledge surfaces in the mind. More detail on the neurobiology of the brain and how it impacts learning can be found in Appendix A. Learning formal systems and the strategies for mapping into and out of formal systems (critical thinking and scientific skills) will also impact the overall shape and connectivity of the knowledge surfaces in the brain.

In summary the brain grows into a mind due to three interacting domains, as shown in FIGURE 36:

·  the physical and biological development of the brain that allows it to learn more complex knowledge structures

·  the psychological training of the brain that guides it to adapt and modify its biological substrate forming habits of mind

·  motivational factors that allow the mind/brain to choose where to apply the limited capacities and energies available for change in creative and learning environments

FIGURE 36. Multimodal domains determine brain state.

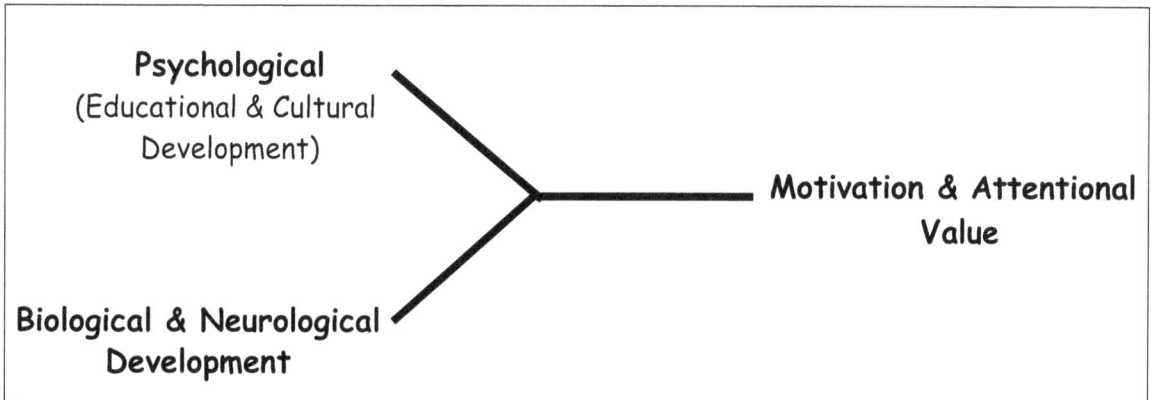

What are the components of becoming literate in science that the brain is required to perform? In order to coordinate the maturing brain capacities with science literacy, an appreciation of the structure of human knowledge is required.

## Summary

We have seen that the nerve circuits of the brain are greatly influenced by prior knowledge or expectation in the interpretation of an observed pattern. These neural circuits always attempt to complete a fuzzy or incomplete pattern in order to resolve ambiguity.

As a consequence, the brain tends to make errors in data analysis and cause-effect analysis. These errors can make it both gullible and predispose it to prejudice. While non-scientific modes of explanation remain susceptible to these innate errors of the human brain, scientific thinking provides a counterweight to these tendencies.

The scientific tools that are used to counter our inherent tendencies toward making errors are the skills of:

· critical analysis,

· skepticism,

· experimental design.

The ability to use these critical thinking skills effectively is one of the most important consequences of an effective education system in our modern civilization. These skills are the language of critical analysis and model-building. Like any language, these skills are best taught starting in the early learning period of a child's development. Furthermore, the learning of these skills must be coordinated with the capacities of the developing brain and the structure of the strategies themselves! How such learning is accomplished is determined by the approach to literacy.

# Unit Three

# Constructing Naive Through Expert Mental Models

# CHAPTER 9

# Teaching the Knowledge Surface

## Information Processing in the Brain

Due to the brain's organization, certain types of information are easily processed while other types of information require complex strategies to be processed. In general, the human brain is superbly designed to work with complex structures, like the appearance of the 1000 species of dogs or the 10s of thousands of faces seen in a lifetime). The brain accurately differentiates between groups and categories, such as dogs versus wolves, or people I know versus people I don't know). No digital computer can do this task done so effortlessly by all humans.

On the other hand, the rote recall of lists of specific data is not a task well suited to the mechanisms of the brain. This is a task which a digital computer is perfectly designed to perform. Lists are well sorted and handled by digital computers because they are comprised from specific data that can be counted.

The human brain processes lists with difficulty. The acquisition and processing of this type of information, when required, is invariably accompanied by strategies for accomplishing those goals. It is the structure of a hierarchy of pattern analyzers, which deal easily with fuzzy or incomplete data that make the brain such an effective system to find general properties of a set of objects. It is the same structure that makes it difficult to handle lists of data. The human brain and a digital computer are constructed very differently.

## Modifying the Knowledge in Teaching

Modifying knowledge surfaces is the task of teaching (and creativity). We can use our graphic model of how the brain builds internal models to represent knowledge structures. We will immediately see that moving a knowledge structure into the internal model space in the brain is the goal of learning. The acquisition of a knowledge surface results in learned behavior. The process of modifying an internal knowledge surface from being simplistic and naive into one that is more sophisticated is a key goal of educational systems and developing disciplinary expertise.

Naive and simple knowledge surfaces are quite flat and unfolded. Increasing sophistication and expertise is captured in much greater folding of the knowledge surface. The simplest form of knowledge is a list. In the Language of Patterns this is a list of *elements*.

The cybernetic sequence tells us that the 'natural-list' is a direct map of each interaction that occurs at a sensor or an array of sensors arranged to respond to the natural world. Each sensor takes a value and these values represent the data. If the data is time dependent a list of values is generated as time moves forward. If the data are a single set of interactions at one time point, the data are either a single value or if there are an array of sensors (like an image on the retina) a list of values each taken at the same time - one value for each sensor. If the data stream is unaltered or simplified in any fashion, the data stream and information stream are identical. A list is now made by mapping the naturally ordered information in sequence, or in the case of a picture as an array or matrix. The knowledge surface in this case is a list because it is not simplified or changed in any way.

## Transforming the Knowledge Surface

Transforming a knowledge surface means folding the list (and remember that human brains are folding machines!).

· Elements: Let's consider a specific list - the integers from 1 to 100.

· Rule: Smallest to Largest

For this discussion the list is going to have structure - it will be organized in an ordinal fashion, i.e. arranged in order from smallest to largest integer. In the Language of Patterns, there are 100 elements, and these elements have a *rule of arrangement*: smallest to largest.

· Background space: List

This stream of data-information is now drawn onto a map, and the data will be arranged on a surface.

The list is a number line and looks like this:

**FIGURE 37.** A number line is a one dimensional list

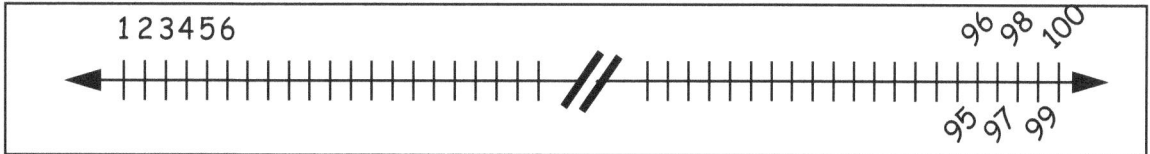

The system description of these ordered, numbered elements is a number line or a one dimensional list. That list can be one row or one column. This knowledge surface can now be used to solve a famous problem:

· Add all the numbers from 1 to 100 together and provide the sum.

The obvious solution is to add each number sequentially:

· 1 + 2 = 3 and then

· (3) + 3 = 6 then

· (6) + 4 = 10 and so on. These actions are seen in figure 38.

**FIGURE 38.** What is happening on the knowledge surface looks like this:

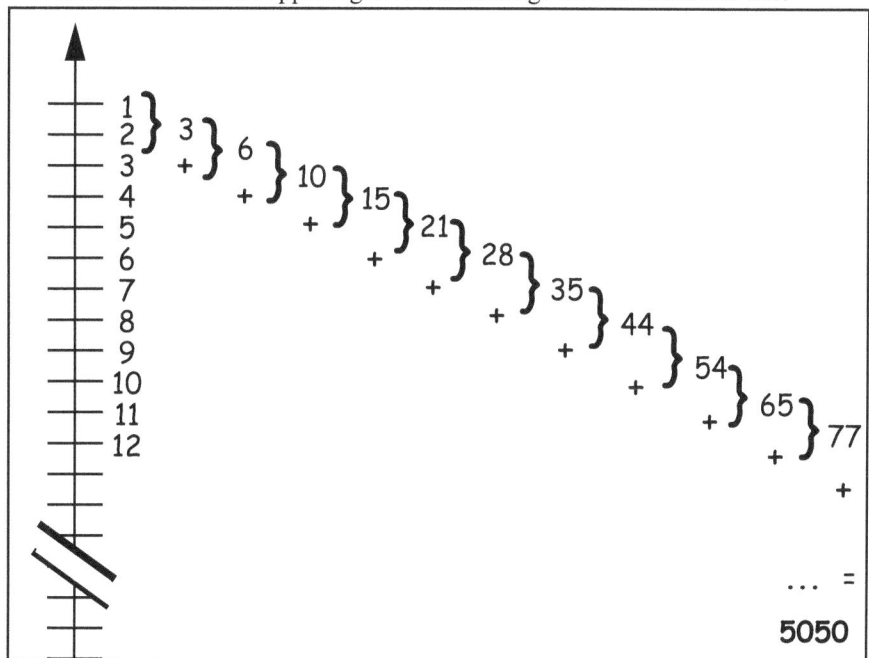

This knowledge-application surface expands to provide the sequential answers needed to provide the solution. In the process it becomes very broad but remains completely flat. It will take 99 addition operations to reach the answer which is **5050**. Very simple but long. For the human brain this is actually intensive, since the data are big and flat. To a computing machine such as the brain that is structured to work on folds and twists, this is a lot to accommodate.

**However**, there is another way to do this problem. A story is toid that the great mathematical genius Leonhard Euler saw this solution as a young student when a teacher gave this time-consuming problem to the class as a way to occupy the students for a long chunk of class time. Figure 39 shows what Euler did.

**FIGURE 39.** Euler's solution was to fold the list.

| 1 |
| 2 |
| 3 |
| 4 |
| 5 |
| 6 |
| 7 |
| ... |
| 95 |
| 96 |
| 97 |
| 98 |
| 99 |
| 100 |

- Start with the list on the left.
- Fold the list in half so that it looks like the list on the right.

Do you see the solution?

- Add each pair of numbers.
- Every pair adds to 101.
- There are 50 of these pairings.
- 50 × 101 = 5050

The number of mathematical operations after folding: 4

- Add 1 row = 101
- Count # of rows = 50
- Multiply 50 × 101
- It takes 4 operations to solve this problem compared to 99.

| | | | | |
|---|---|---|---|---|
| 1 | + | 100 | = | **101** |
| 2 | + | 99 | = | **101** |
| 3 | + | 98 | = | **101** |
| 4 | + | 97 | = | **101** |
| 5 | + | 96 | = | ... |
| 6 | + | 95 | = | |
| 7 | + | 94 | = | |
| ... | | ... | ... | |
| 45 | + | 56 | = | |
| 46 | + | 55 | = | |
| 47 | + | 54 | = | |
| 48 | + | 53 | = | |
| 49 | + | 52 | = | |
| 50 | + | 51 | = | |

Folding the knowledge surface makes the solution much easier and faster. This is an example of a more sophisticated knowledge surface. It is now folded and no longer flat and simple. This is the extent of the complexity to which we will take this discussion. It is accurate to regard the flat knowledge surface as naive - needing extensive exploration and concrete detailed search to find useful knowledge. The folded surface by its nature brings useful and salient knowledge to bear more easily, i.e., it is more like an expert in its action.

Folded surfaces are more sophisticated or expert compared to flatter surfaces that are concrete and naive. As discussed earlier in Unit 2, the neurobiology of the brain, because of its cells and connections, is designed to work by folding information and re-folding knowledge surfaces.

- **Learning** is the name we give to the process of folding and construction of the folded surfaces.

- **Teaching** is the method of knowing the surface with which we start, and then catalyzing its folding from simple and naive listings into more complex expert systems.

Here is a notional example of how the letters that form the word **flower** can be found from the list of the alphabet.

**FIGURE 40.** Flat and planar surfaces with long interaction distances and uncorrelated relationships are slow and inefficient.

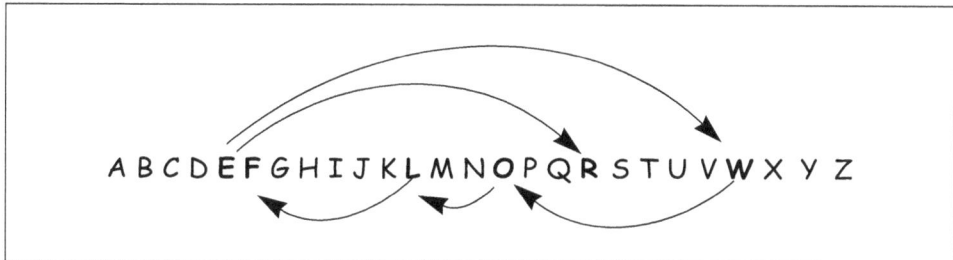

The knowledge that "F" is often associated with "L" or "R" and that "E" is frequently part of "ER" and "OW" are typically found together might generate a more sophisticated knowledge surface for an early reader that looks like this.

**FIGURE 41.** Folded surfaces are highly correlated, have fast interactions and are fast and efficient.

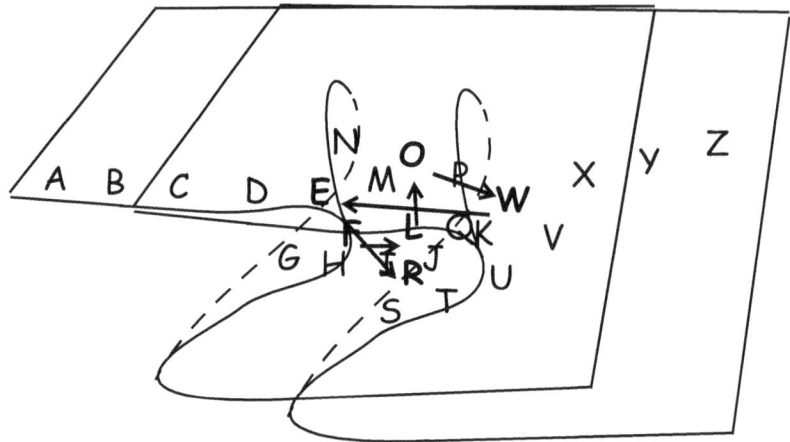

## Effective Teaching

Teaching is a constructivist process of turning naive internal knowledge surfaces into more complex expert-like surfaces. How does a sophisticated expert system catalyze a simple flat naive system to fold in an organized fashion? The answer is by interacting with it. This is not as simple as bringing the naive surface into proximity with the expert surface and expecting magic to happen. Folded surfaces do not fit with flat surfaces easily. If they do not fit they can not interact.

Everyone has had the experience of being a naive or early learner and meeting an expert who "talks over their head". Another way of saying this is the low dimension or flat knowledge surface can not interact with most of the higher dimension knowledge that resides above and below the flatter surface.

Using our simple example, figure 42 shows what the interaction of these two knowledge surfaces looks like:

**FIGURE 42.** A 2-dimensional planar surface does not interact readily with a 3-dimensional geometric shape.

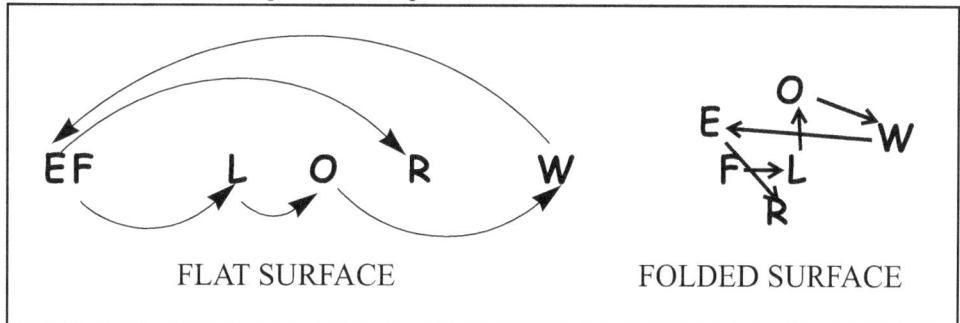

FLAT SURFACE                    FOLDED SURFACE

The naive mind regarding "flower" in a flat dimension can not imagine how the expert mind puts the same pieces together in a 3-dimensional space. At the same time the expert mind does not regard it's folded knowledge-laden structure as anything but simple, obvious, and easy to use.

Teaching requires that the person possessing the expert internal model can recognize that the naive internal model is flat and concrete - filled with list knowledge. The goal is to help the flat surface fold itself and become more sophisticated. In ideal teaching the complicated knowledge structure first unfolds allowing the flatter naive surface to engage it. This is engaging the student where they are. Then the expert knowledge surface guides a reformatting of the naive surface into a more folded and thus sophisticated surface.

What a surface can be reformed into depends to a large extent on what it is at the start of the process. Recall that Ausubel said that the single most important factor in learning is what a student already knows when they come into class. The practical process of catalyzing knowledge surface development will be described in the following chapter.

Finally, consider a teaching process with our knowledge surface model. The naive student knowledge surface is simple and relatively unfolded. Creating and then modifying the student internal knowledge surface is going to take work. Therefore, the first step in any teaching situation will be to gain the student's interest and motivation to do that work.

The FIGURES 43, 44 and 45 that follow demonstrate the changing of the knowledge surface with effective teaching. The first and hardest work is mapping a list into the folding machine that is the brain. Such a list looks like A, which is mapped onto the naive mind, B.

**FIGURE 43.** Flat knowledge surface of the naive, early learner.

**A**  A B C D E F G H I J K L M N O P Q R S T U V W X Y Z

**Naive list**

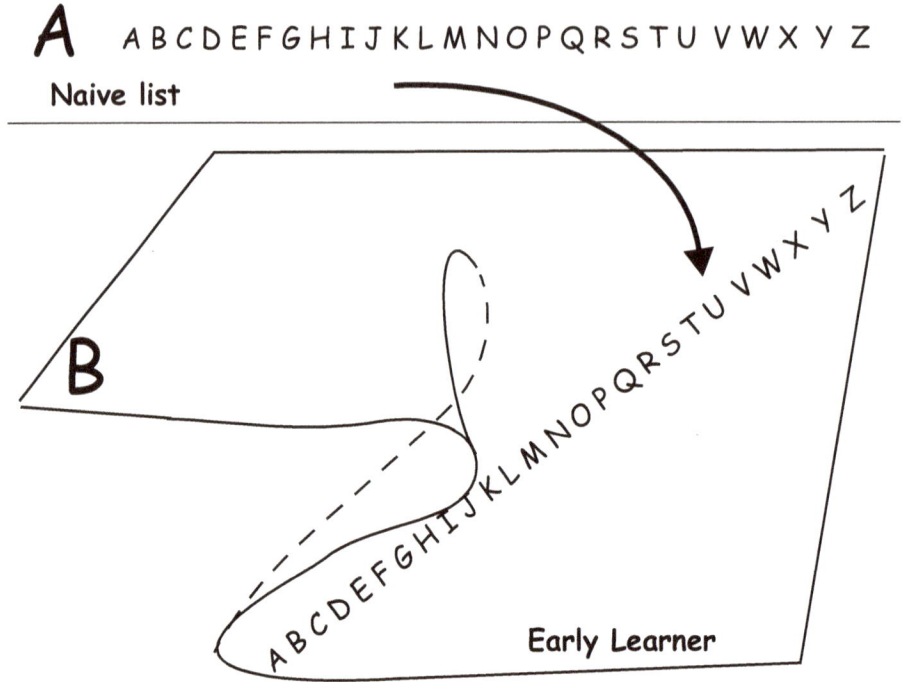

B

Early Learner

Our teaching objective is to ultimately generate a knowledge surface that looks like this:

**FIGURE 44.** Highly folded and complex expert knowledge surface.

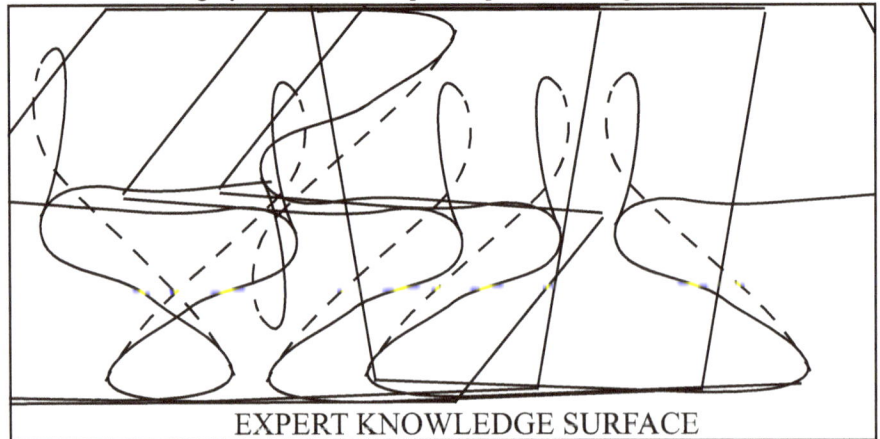

EXPERT KNOWLEDGE SURFACE

In order to do this the knowledge structures must interact. To maximize interaction the teacher elicits a model of the student mind by engaging them and eliciting this knowledge state from the student.

FIGURE 45. First the expert opens to accommodate the student. Then the student surface is engaged and catalyzed to fold in the expert direction.

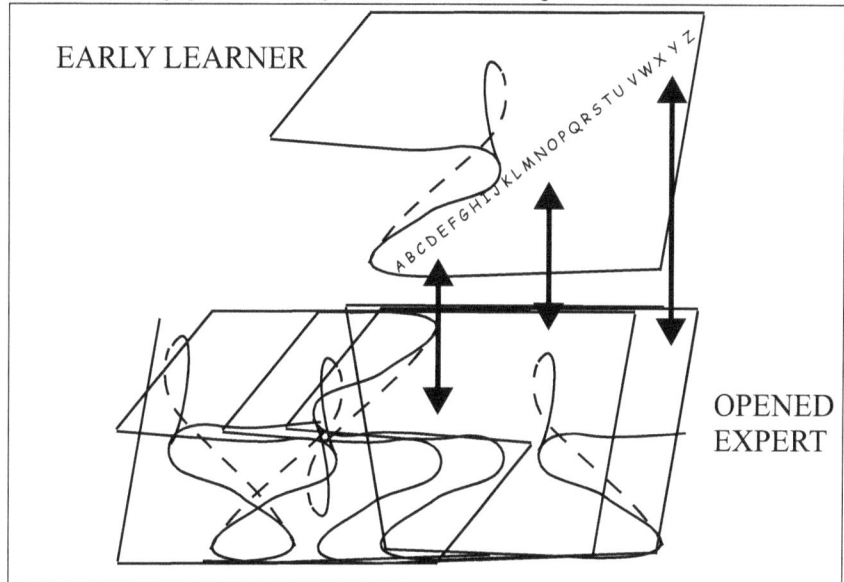

The expert teacher operates at the student level first. The expert knowledge surface is opened so the surface of the teacher and student can interact and begin the process of helping the student discover in a structured path the new relationships.

The student internal model is then encouraged to fold into a more sophisticated surface that contains the internal knowledge that increases connectivity of information and decreases the cost of finding and applying the process. At the end of the learning cycle, the student's internal model has become more expert-like and a degree of learning has occurred.

Measuring that the learning is occurring is the job of formative assessment. Measuring that learning has occurred is the job of summative assessment. Measuring that learning is meaningful and can be applied to new situations is the job of authentic assessment. The formal steps to constructing a series of learning cycles will be described. in the next chapter.

# CHAPTER 10

# Assessment Framework Design Model

The Assessment Framework Design Model (AFDM) is the method developed by and used throughout the SymmetryScience™ curriculum to implement the Cycle of Pedagogy to teach science. This approach has been created from a merger of best practices from cognitive neuroscience, curriculum design, educational research, and the emerging field of educational neuroscience.

## The AFDM in Outline

1.  The AFDM treats learning is a constructivist process [as opposed to behaviorist] and generally requires a learning cycle to implement the learning

2.  Constructivism is intrinsically a modeling process and therefore requires systems analysis and design.

3.  Learning cycles require assessment processes as a the framework on which learning is monitored and refined dynamically for each learner.

4.  The merging of systems analysis/design, cognitive neuroscience measurement, and learning cycle processes gives rise to the AFDM.

## Vocabulary

A quick preview of some vocabulary used in this chapter is valuable.

**Behaviorism** - this is the view that it is impossible and pointless to seek knowledge about the internal workings of the brain or mind and that only the stimulus and response pattern of an animal is appropriate for psychological study. Behaviorism focuses only on observable behaviors. Behaviorism holds the general view that the nervous system is a reticular network that acts by aggregate or mass action. This means essentially that the biological brain is a nerve network with all parts being equal and non-distinct. Edward Thorndike, B.F. Skinner, and Karl Lashley are major figures in behaviorism. Most of these ideas (though still popular especially in educational and social science spheres) are not well supported by the detailed experimental evidence of the last 70 years.

**Cognitive Neuroscience** - the merger of cognitive psychology, neurology, computational neuroscience, and the broader research approaches of neuroscience has formed the current field of cognitive neuroscience - a field that seeks a biological, chemical, and biophysical understanding of the mental processes that underlie how people think, learn, know, and remember. Cognitive neuroscience is largely based on the biologically supported view of structuralism in which the specific anatomical parts of the brain do specific tasks. The interaction of the parts leads to emergent behaviors. The empirical evidence from the last 70 years in neuroscience research strongly supports these constructs.

**Cognitive Psychology** - the study of mental processes: how do people perceive, think, learn, and remember. Cognitive psychology is concerned with internal states of mind and the processes that lead to these internal states. Constructivist views of internal mental states are the common world-view of cognitive psychology and neuroscience. Cognitive psychology has largely merged into the field of cognitive neuroscience.

**Constructivism** - this view of learning is based on the idea that the learner acquires knowledge of the world through the construction of internal models. Thus experience is something that the learner assembles into a meaningful internal model or schema. The focus of constructivist teaching theory is on methods to coordinate both internal states and external experience to lead to effective learning.

**Education** - the process of providing or receiving formal instruction - usually at a school or university setting.

**Educational Psychology** - largely the subject of school psychology, the study of how schooling and formal instruction is influenced by social, personal, and cognitive processes. The field includes psychometric assessment of students and the study of the effectiveness of educational or pedagogical approaches to learning.

**Learning** - a relatively permanent change in behavior caused by experience (stimulus). In behaviorism learning occurs by conditioning processes (classical, operant, or observational). In cognitive neuroscience/psychological paradigms learning is constructivist in nature.

**Pedagogy** - the method and practice of teaching, especially as an academic subject or theoretical concept.

# AFDM Terms

These will be defined in the following discussion:

- Performance Objective
- Learning Elements
- Learning Trajectory
- Formative Assessment
- Summative Assessment
- Authentic Assessment
- Learning Cycle (the 5 E's)
- Knowledge Analysis (3 C's+P)

Learning is a constructivist process and requires a learning cycle to implement the learning. This is in contrast to a behaviorist approach that uses a reinforcement schedule to implement learning.

The primary lesson from cognitive neuroscience experimental data supports a "constructivist" approach to learning. Constructivism requires that understanding the process of learning means recognizing the internal state of the brain and mind and specifically acquiring knowledge about how those internal states are attained and altered. Therefore, learning is a function of both the individual and the environment. To teach yourself or a student new elements comprising knowledge is to foster adoption of models consistent with expert knowledge; this implies a concurrent reorganization of many components: concepts, principles, conditions of applications, links between real world elements and elements in the model, as well as beliefs and assumptions about what should be examined. Such reorganization can be viewed as moving from one state of equilibrium (the initial model) to another (the more advanced model), overcoming erroneous assumptions, misconceptions, and bias along the way.

## Constructivist Processes

The formulation that the brain constructs an internal model of the world is a well-developed and useful theory with applications in neurology, cognitive neuroscience, education, and the social sciences. There is substantial experimental evidence that the nervous system employs a computational strategy in which the process of feature abstraction processes sensory data from the level of stimulation to the level of cognition. The role of the generated internal models is central to the normal function of the brain. Learning, at least at the cognitive levels above simple adaptive behavior (where behavioral approaches can be quite practical), may be usefully regarded as the process leading to alteration (revision, reconstruction, or embellishment) of these internal models. Learning is a function of both the biological substrate and the environment. If educational practice can be related to the construction and revision of these internal models in a coherent, theoretical formulation, the underlying biological and psychological processes can be connected to the behaviors that are manifested by the learner.

## Learning Cycle

A learning cycle (e.g. the 5 E's) is the term used to describe the steps in a pedagogical paradigm that guides construction of internal models. A learning cycle coordinates teaching steps with cognitive neuroscientific understanding of learning:

- **Engage** - focuses attention and attaching emotional salience to a learning event.

- **Elicit** - organizes a description (system model) of the current internal model which allows the teacher to connect what is already known to the topic under study.

- **Explore and Explain** - recognizes the general drive of the human brain to find correlations and causal chains in data and experience which is the explanatory modeling process.

- **Elaborate** - uses reflective and metacognitive (critical analysis of what is known) processing to construct models.

- **Evaluate** - recognizes the essential role of assessment to the process of reflective and responsive constructivist learning.

The SymmetryScience™ Method uses all of the 5 E's in each learning activity.

## Internal Models

Internal models are systems of knowledge. The Language of Patterns is a metacognitive or critical analysis tool that is used to describe and detail systems and models. Models are partial descriptions (or abstractions) of systems of interest. A "good" model is an abstraction at an appropriate level of detail that 'accurately' represents the reality of the system of interest. A formal definition of a system can be given. A system is a set of connected things or parts that form a complex whole. The properties of systems that are measurable are called **observables** and are typically emergent from the overall operation of the system. A systems analysis includes a notation of its **elements** (parts, components, events, or things), the relationship **rules** relating the elements, the **context (or background space)** in which the elements and rules are found and operate, and the **observable (emergent) properties** that measured together define the state of the system. Typically the elements comprising a system are themselves further describable as systems, that is, they are sub-systems.

## Knowledge Analysis

In the context of the AFDM the structure of a knowledge system can be described with a systems approach (the 3 C's + P).

1. **Content** maps the elements of the knowledge system.

2. **Concept** map the relationship rules of the knowledge system.

3. **Context** maps the relationship background context to the knowledge system.

4. **Process** maps the tools and skills needed in the discipline. Processes are also technically systems with their own elements, rules, and background space for performance. However, for all disciplines, process and methods are often so foundational that the AFDM functionally identifies them as a special type of knowledge system.

It is worth recognizing that the model phases represented in the cycle of pedagogy (brain, knowledge, pedagogy) are themselves multidimensional. For example, the neural structure building mental models (brain to mind) can usefully be described along a developmental axis (brains are structurally different in a dynamic fashion from infant to old age) and a psychological axis (knowledge development builds from naïve to expert based on experience, learning, and education).

## A Very Brief History Of Learning

Following the early to mid 19th century most neuroscientists viewed the brain as organized around specific structures. This **structuralism** was the dominant view that actually predated the ability to see the cells and connections of the nervous system under the microscope. The scientists who dominated the debate were Franz J. Gall and Johann Spurzheim, and then Paul Broca and Carl Wernicke. In the later part of the 19th century (still without the ability to visualize the nervous system empirically), the notion of aggregate or mass action and a reticular or network structure of the brain came to dominate. These ideas were consistent with the social ideas of the time that were turning toward the concepts of collective action. During the first half of the 20th Century the behaviorist school of thinking came to dominate the ideas of learning, memory, and the highest of cognitive functions, including the acquisition of language. **Behaviorism** dominated psychological study and the ideas of the learning process. Three major types of learning are described in the behavioral view: classical conditioning, operant conditioning, and observational learning.

## Behaviorist Models of Learning

In classical conditioning learning occurs when an association is made between a previously neutral stimulus and a stimulus that naturally evokes a response, i.e. Pavlovian learning. Here the external environment acts on the animal whose nervous system makes the correlation between events: stimulus and response.

Alternatively, learning via an operant conditioning process occurs when the probability of response occurring is increased or decreased due to reinforcement or punishment. Here an animal emits a behavior into the environment and discovers whether and how it is rewarded. It operates on the environment. The scientists Edward Thorndike and B.F. Skinner developed operant conditioning to a high ideal with the central concept that the consequences of our actions shape voluntary behavior.

Finally, observational learning was a type of learning that occurs through the observation and imitation of others. This was largely the work of Albert Bandura who demonstrated that people will imitate the actions of others without direct reinforcement. Effective observational learning requires attention, motor skills, motivation, and memory.

In the behaviorist approach, these output behaviors of learned actions did not consider how the internal state of the brain or mind accomplished these new stable states of behavior. In the later half of the twentieth century the demonstration of actual limits of working memory by Earl Miller and the demonstration by Noam Chomsky that natural language could never be learned simply by operant conditioning (requiring an internal computational structure to extract language from experience - neurolinguisitics) lead to the falling out of favor of behaviorism as a useful approach to understanding the cognitive actions of the nervous system.

Bandura combined observational learning with socially influenced learning concepts. He suggested that live models, verbal instructions (descriptions and explanations of behaviors), symbolic models, and intrinsic mental states were important to learning. Thus the ideas of modeling actions and building internal models, as well as the recognition that learning did not always occur or change behaviors, lead to a view that an internal modeling process was occurring in the brain (mind). Certain processes were required in the modeling process: attention, retention, reproduction (practice) of the behavior, and motivation to attempt to model the new behavior.

Cognitive psychology is probably reasonably considered the child of Jean Piaget who studied and organized the ideas that the intellectual development of children into adulthood was built around an ontogeny (sequence of states) of knowledge handling and development that was intrinsic and controlled by biology more than just experience. Thus experience is processed by cognitive processes and abilities, rather than generating those processes and abilities via experience. We will not recapitulate the ideas of Piaget here except to note that he identified a series of stages in which the developing child seeks to build internal models of knowledge about the world (called schemas) in which new information and experience is first assimilated and then accommodated by the internal schema, forcing the internal model to change. The change occurs via a process of equilibration in which the new information is balanced with the older structure of knowledge leading to movement between stages of thinking.

## Educational Neuroscience and Research

Educational neuroscience is a research-based practice. The ideas of Piaget have evolved to a great degree, but the fundamental ideas that learning should be aimed at levels for which the learner is developmentally ready and that internal models are dynamic and fluid constructions is still the research view of much of the work in cognitive neuroscience. There is strong experimental support for constructivist teaching strategies in which students are provided supportive, active learning environments, social, cooperative and peer teaching opportunities, and reflective learning. When well executed, these approaches are effective in helping the learner discover the fallacies and inconsistencies in their thinking and to modify them.

Neuroscience and cognitive psychology provide an organizational view of the nervous system as a biologically deterministic system designed to respond to its environment with learning actions that are a consequence of internal organization of knowledge into models of the outside world. This view has had a profound impact on formal learning strategies and theory and forms the basis of the emerging field of Educational Neuroscience.

Cognitive neuroscience accepts the context of constructivism based on the empirical evidence and therefore explores how internal models of knowledge (and action) are created and updated. In its application to learning and education, this world-view supports and explores certain types of instructional strategies that are therefore reflected in certain pedagogical approaches.

Educational neuroscience is a scientific endeavor. Demonstrating that a strategy actually works in the classroom or individual learning environment requires a scientific approach to educational research.

Educational research requires that a teaching strategy (an intervention) meets several criteria that are tied to the definition of learning (a relatively permanent change in behavior as a consequence of a stimulus):

1. **Effectiveness** - How well did the intervention lead to proximate learning?

2. **Durability** of the effect - How "permanent" is the learning effect?

3. **Transferability** - This is especially important in higher cognitive learning where the purpose may be to generalize a process (diagnostic problem solving) to other similar or analogous categories. It can also be important in the obverse since differentiability or limiting of the cases to which something learned is applied can be just as important if the

generalization might lead to misconception. In other words preventing behaviors that are captured in the following: "If all you have is a hammer, everything looks like a nail."

## Teaching Strategies

The research from cognitive neuroscience demonstrates the effectiveness of the following learning and teaching strategies:

1.  The incorporation of a "Learning Cycle".

2.  Individual discovery and constructivism mixed with social or structured discovery.

3.  Reflective learning strategies. "Tell me what you are learning and how you understand it."

4.  Cooperative learning strategies. "Let's learn this together and be able to discuss what we now and know and are learning."

The SymmetryScience™ curriculum incorporates all of these principles.

## Teaching Using the AFDM

The Assessment Framework Design Model (AFDM) provides a coherent approach to lesson/course or curriculum/program development. It can be easily adapted to any learning situation whether for a short single explanation (very hard to do well) or for longer term educational interactions such as a structured curriculum. It is the paradigm that executes the Cycle of Pedagogy.

The AFDM guides movement from an initial level (or state) of knowledge to a final point that must be identified. The ultimate goal of a teaching sequence is called the performance objective. In order to know where you are going and to assess that you have successfully arrived there requires a statement of the state of knowledge at that performance-objective point. The parts of the AFDM include:

1.  **Performance Objective**: Overall outcome measure that reflects how the knowledge and new abilities will show. How will the student be changed (what behavior is altered long term)?

2. **Learning Elements** (Objectives): What new knowledge or added content, concepts, and context students are expected to know or do in order to achieve the performance objective. What are the knowledge pieces that need to be taught in order to achieve the performance objective?

3. **Learning Path** or Trajectory: How you will arrange individual pieces of information (as content, concepts, context, or process or the 3 C's+P) built from the 5 E's (below) to achieve the learning and overall performance objectives.

4. **Assessment**: What evidence is used to judge that the learner has achieved the desired objectives (you explicitly define these as needed). Assessment is considered as a crucial observable that confirms your alignment to your learning path. It must not be an afterthought but must be integral to the educational process.

**Example**  In order to make this AFDM process more specific, let's see how the AFDM is used to guide the design of a teaching path. The metaphor of a piece of jewelry is useful.

1. The "**Performance Objective**" is what your overall teaching structure will accomplish. Sometimes called the "indicator of learning" it is a statement of the outcome behavior generated by the new knowledge or abilities that you have taught. In a sense this is a statement: "I am going to make a piece of jewelry to wear around the neck [a necklace]" vs. " I am going to make a piece of jewelry to wear around the wrist [a bracelet]". The performance objective is an emergent property or behavior of the system of learning that you are constructing.

2. The "**Learning Elements**" are the actual new pieces of information, skills, perspectives, or concepts that need to be mastered in order to reach the performance objective. Implicit in this are the assumptions of what the student already must have mastered and will bring to the learning environment. These objectives are the 3 C's + P (Content, Concepts, Context & Processes/methods) of the knowledge structure that you must craft with the student. You should be able to explicitly list both what the student must know a priori and what you will add to the student's knowledge base in your teaching. These "learning objectives" are like the beads, string, and clasps that you

would need to assemble in order to build your piece of jewelry. Each learning objective or element is a subsystem comprised of content, concepts and context. These subsystems will be linked together to build the knowledge system that will give rise to the "emergent behavior" that is the "performance objective". Though most teachers use the term "learning objective", the term learning element reminds us of the systems nature and connectivity of the bits of knowledge that lead to the performance objective. Using the term "element" preferentially is a reminder that the Language of Patterns provides the analysis needed to systemically construct the learning elements.

3.    The "**Learning Path or Trajectory**" is how the learning objectives are put together to lead the student to an operational knowledge of the topic. You organize the learning objectives in the teaching space to help the students "unfold" that knowledge structure that you want to give rise to the "performance objective or net behavior". Learning trajectories organize the teaching elements and can take a variety of forms:

**FIGURE 46.** The five prototypical learning trajectories in AFDM learning design.

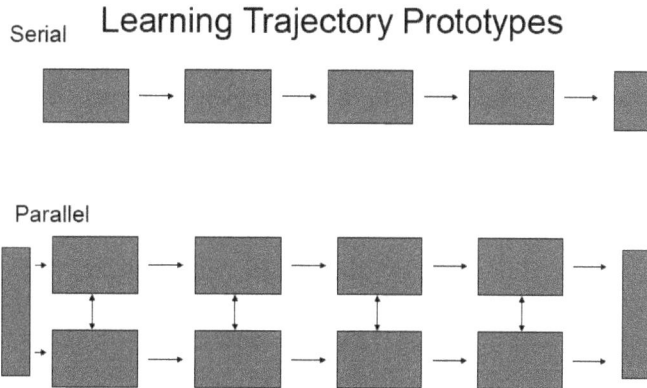

Learning Trajectory Prototypes

Serial

Parallel

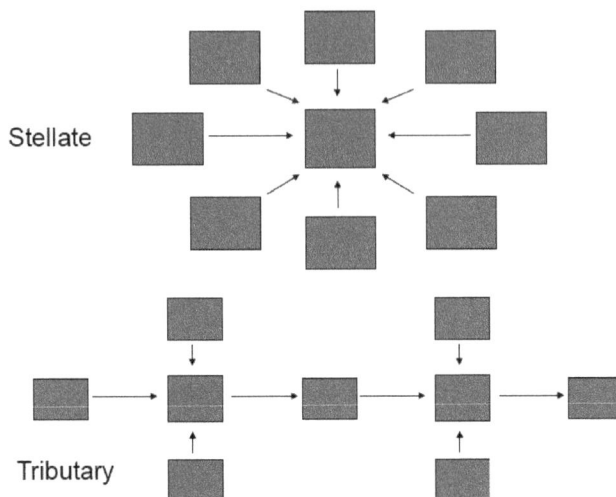

Stellate

Tributary

Remember that each element along the path to the performance objective should be constructed from a learning cycle:

4.    A **Learning Cycle** includes the activities or opportunities for learning are each a learning cycle. The learning cycle is broken into 5 components:

- **Engage**: Get the learner's interest and buy in to attend to the learning opportunity.
- **Elicit**: Discover what the learner knows. Asubel said that the single most important factor in learning is where the student starts and what they already know.
- **Explore and Explain**: Explain the topic at the learner's level.
- **Elaborate**: Extend the understanding by the student of the topic toward the level of understanding that is required to achieve the performance objective.
- **Evaluate**: Measure what is being learned. Has the performance objective been reached?

**5.** The Assessment Tools are developed and deployed throughout every lesson/curriculum and should be characterized by all of the following:

- **Formative Assessment** - How will you know the student is with you or following on the path that you are laying out in the lesson? What "dipstick" measures will you use in real time? "Dipstick" measures gauge the student's understanding along the learning cycle:

  - "Are you with me?"

- **Summative Assessment** - Tests vocabulary, facts, and techniques of the knowledge structure to be taught. It measures the acquisition of knowledge or competencies planned in the 3 C's+P.

  - "Can you tell me what you learned"?

- **Authentic Assessment** - What real world test shows that the student has adopted a new or changed behavior. Can the student incorporate the new knowledge into their internal model of the topic and apply it outside the teaching environment? It is a real world measurement demonstrating changed behavior attributable to incorporating new knowledge or competencies. Often the assessment looks at how the new knowledge can be applied to a different situation.

  - "Show me how what you have learned applies to this?"

# Unit Four

# Knowledge Is Structured

# CHAPTER 11

# The Structure of Human Knowledge

## Models and Formal Systems

Knowledge of the world is gained when the brain builds a model of the world!  The world we experience includes the natural world and the imaginative world. Experience is gained through observation, participation, and imagination. Personal knowledge exists as a model in the brain.  However, human knowledge is not limited to personal experience.

Because the human mind can learn from others, knowledge is accessible through the social enterprises of the community. The biological human is a strongly social organism bound by community and culture. Language, habit, and belief are key emergent properties of human social interactions. The key to social knowledge is through a shared formal world. When personal knowledge is shared or integrated on a social and societal level, the internal models are modified and merged.

## Experienced World

The experienced world is detailed, specific, diverse, and non-uniform.  It can be experienced and modeled without education or enculturation. The mind can map experiences onto a formal world. Formal worlds are unified, abstract, and coded. Education is required to provide the codes. The mapping of experience onto a formal system and communication with others requires critical thinking strategies and skills.  These skills are refined with an education.

These knowledge structures or models are representations of the world.  They are built in a "virtual" world called a formal system.

When the brain regards the natural/real world, it extracts certain observable features through its senses. These features are then translated into a model that exists only in the brain. Such systems of representation are called *formal systems* and include:

- the symbolic languages of mathematics
- spoken and written language
- musical composition
- chemical notation
- graphical art
- theatrical and cinematographic art
- economic and business accounting
- legal analysis and law
- many others as well

## History of Formal Systems

Why did human cultures develop formal systems? What was the advantage? As an example we will consider the history of written language.

Early civilizations collected their experiences and cultural knowledge and passed it on to the next generation by story and song. Initially, there was no written record of the knowledge of a culture.

The earliest records of knowledge were made by creating objects or pictures that were graphic replicas of events and objects. These representations, while not yet a written language, are actually objects of the natural and experienced world. There is no significant formal aspect to these representations.

FIGURE 47. Pictures (above), pictographs and ideographs (following) predated languge as representations of knowledge.

As civilizations accumulated more knowledge, the method of stories and sagas were unable to transfer all of the information. Methods of recording the information systematically were required. The earliest forms of written language were hieroglyphic **pictographs**. These are recognizable models of real objects and events, much like

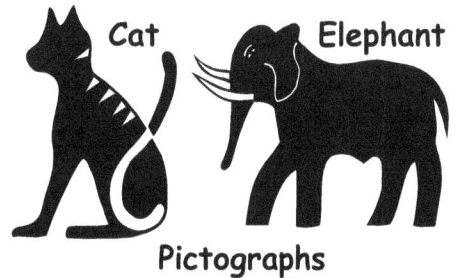

**Pictographs**

the primary sensory model of Rover. This type of representation is much like the discrete mapping made to a knowledge surface in early learning. Pictographs were useful for representing objects but were limited in the information that they could transmit.

The next stage in the development of written language was the incorporation of more abstract concepts, such as ideas and functions. Pictographs evolved into **ideographs**. The meaning of an ideograph depended on its context and usage, rather than just on the actual physical object from which it was modeled. The picture became more symbolic and the ideas that it transmitted became more abstract (fewer detailed, recognizable features and more general meaning). The ideograph is similar to the broad category region formed in a

be
neglectful

catch
fish

**Ideographs**

knowledge surface, as many specific examples are linked into a more general concept space (Chapter 7).

The formation of ideographs requires precise mapping of the forms comprising each one. This ensures that they retain their information content. To train students in the precise use of these forms, "formal" education is required in a civilization. Ideographs, though formal characters, do not have the full abstract capacity of an alphabet. Since ideographs still carry substantial concrete representational information, their use requires a very large set of specific characters.

An alphabet is the ultimate formal system.

Alphabets use a very limited number of forms (letters), that can be arranged in an infinite variety of structures (words). These words can be arranged according to general rules (syntax) to create a vast set of sentences, paragraphs, and arrangements of words. Depending on the space in which the words are arranged, alphabets can form essays, books, poems, graffiti, dramatic works, prayers, just to name a few.

**FIGURE 48.** With an alphabet a code can be made to represent everything and anything

## Formal World

A formal system is efficient; its power is derived from the huge diversity of structures built on the foundation of a few simple forms and rules. Formal systems are codes. There is nothing about the word or letter itself that carries any information about the original object or idea. Achieving literacy (appreciation of the unity of forms and the diversity of their many structures) requires specific formal instruction to provide the code-breaking knowledge required to understand and use the formal system. Formal systems, though created entirely by the human mind, are mapped and computed just like the knowledge surfaces that contain the objects of the natural world.

The elements, rules, and space of a formal system are internally consistent. Formal systems may or may not have a clear correspondence to the elements, rules, and space of the natural world.

For example, in making an accurate representation of the properties of a metal bar musical notation is not useful. Mathematics and chemical notation are better choices for such a model.

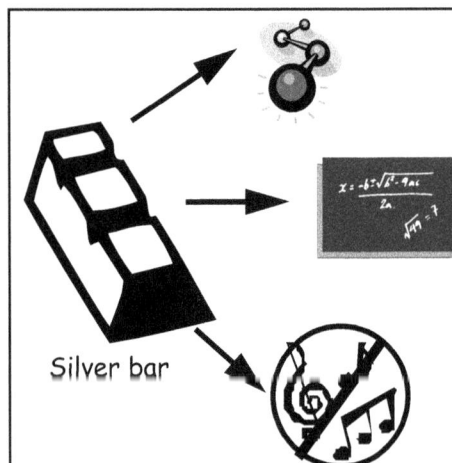

Silver bar

**FIGURE 49.** Ideal formal systems choose the best model of representation to capture the real system.

The world of knowledge includes the natural world and the formal world of the mind. The natural world exists independently of the humans who explore it. It is important to recognize that humans are a part of that natural world and are subject to its rules independent of humanity's viewpoints. The formal worlds of the mind have their own internal rules. These are human constructions and may or may not have a relationship to the natural world.

These two worlds are usually well appreciated. The natural world is explored by observation, wonder, and need. The formal worlds are taught in schools, preparing the learner with a formal education. A formal education provides the code book and rules of the formal world. Both worlds map information onto knowledge surfaces in the brain.

## Region of Critical Analysis

An important third world also exists - a region the processes between the natural and formal worlds are mapped. This area focuses on critical thinking, modeling, and analysis.

The process of assigning a representation of the natural world in the brain (onto a knowledge surface) is like making a map. Making a map is a model-building process in which a feature of one system is transformed by a certain rule into a feature in the second system. The choice of the mapping process and the formal system is part of the critical thinking process.

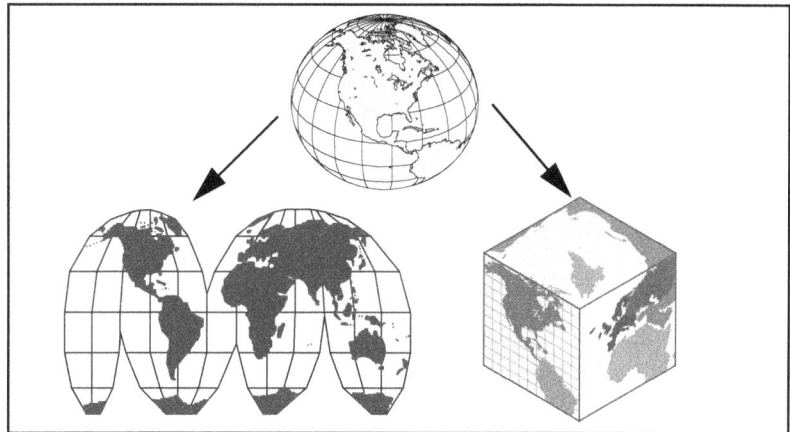

**FIGURE 50.** Accurate and practical mapping is an essential component of critical analysis.

**FIGURE 51.** Critical analysis is defined by systems mapping

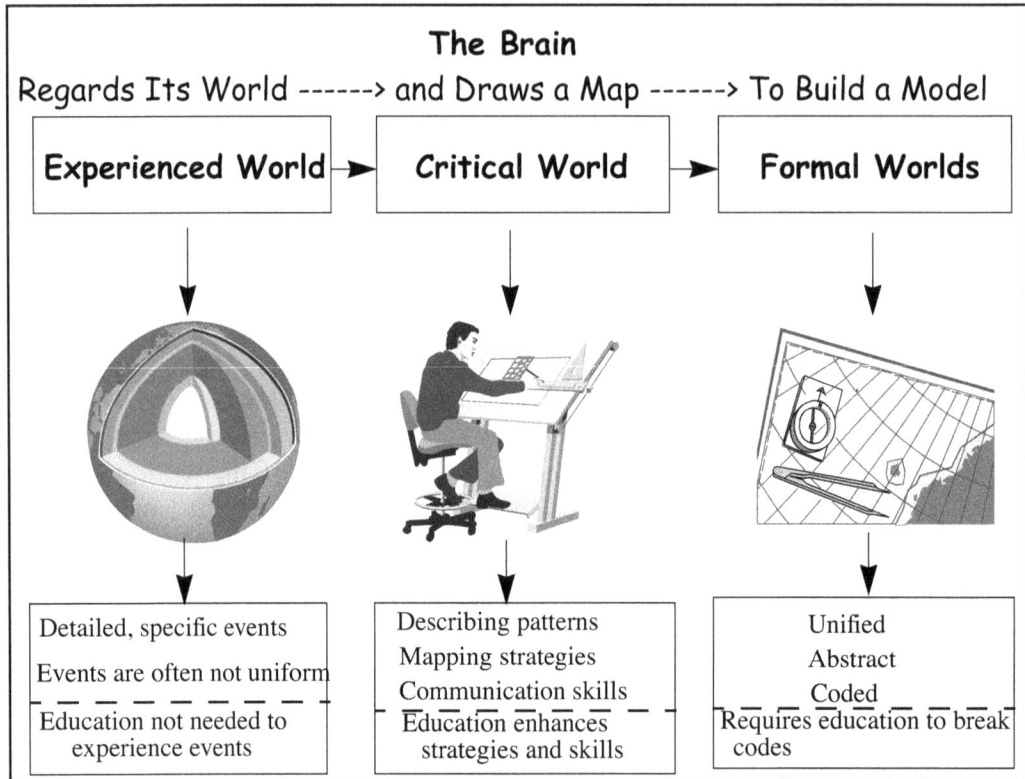

## The Brain
## Regards Its World ------> and Draws a Map ------> To Build a Model

| Experienced World | Critical World | Formal Worlds |
|---|---|---|

| Detailed, specific events | Describing patterns | Unified |
| Events are often not uniform | Mapping strategies | Abstract |
| | Communication skills | Coded |
| Education not needed to experience events | Education enhances strategies and skills | Requires education to break codes |

In summary there are three regions in knowledge construction:

- the experienced region which is most concerned with the natural world and is governed by the rules of the universe; and the imaginative space which represents our artistic sensibilities.

- the formal region with rules created by humans

- the critical analysis region which interfaces the natural pattern features with formal pattern features by a process of critical thinking and analysis

FIGURE 52. Formal or critical strategies are used to map features of the natural world into an internal model

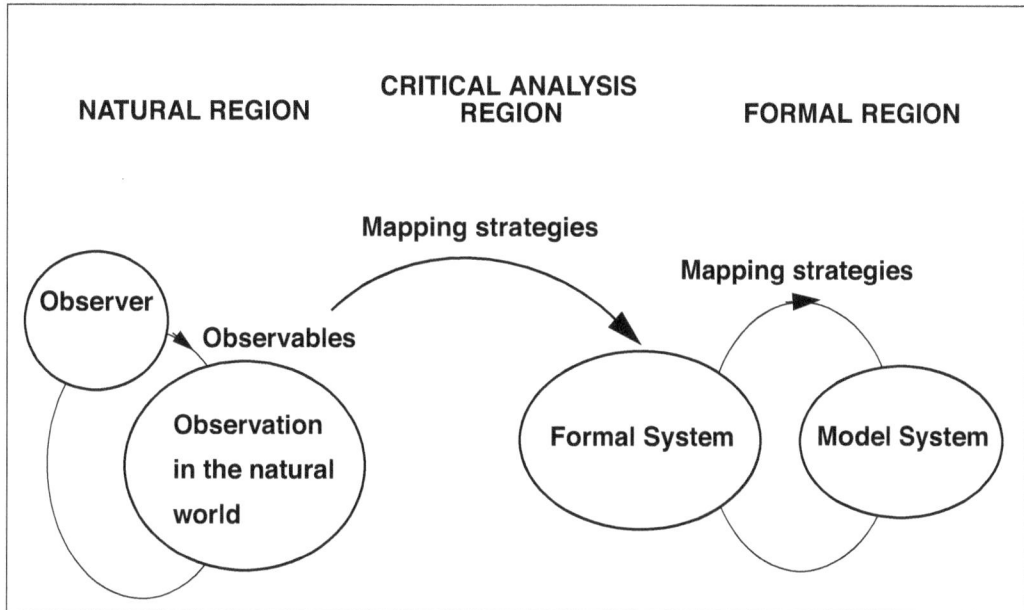

FIGURE 52 shows the internal process that is contained within each of the modeling domains designated in the Progression of Inquiry. Whether constructing an observational, explanatory, or experimental model, this critical modeling process is required for each domains. As we will learn in the next chapter, this critical analysis cycle applies to all human knowledge building - not just scientific inquiry. We will see that the critical difference is the approach to verification and validation of the models that are built. But in most cases the approach to critical analysis whether scientific or non-scientific is identical.

## Five Steps to Modeling and Critical Analysis

The process of acquiring knowledge about systems through modeling and critical analysis may be organized into five steps.

FIGURE 53. A simple set of diagrams can guide this discussion.

**Step 1. Observation**

Observations (shaded in gray on FIGURE 53a) are made in the natural world by the human senses. Tools (such as a microscope or weight scale) are often used to extend the senses.

The pattern features of observed objects and events (the properties of systems) are detected (observables).

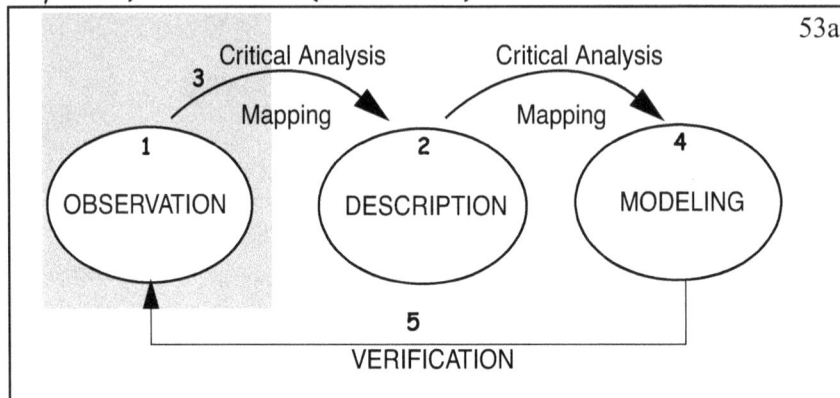

53a

**Step 2. Description** The description (shaded in gray on FIGURE 53b) of pattern features is an important critical thinking skill, because it enables recording, discussing, and debating with others about the observation. The Language of Patterns can be used to describe the pattern features of the system in terms of its properties, elements, rules that relate the elements to each other, and background space in which the system being observed exists.

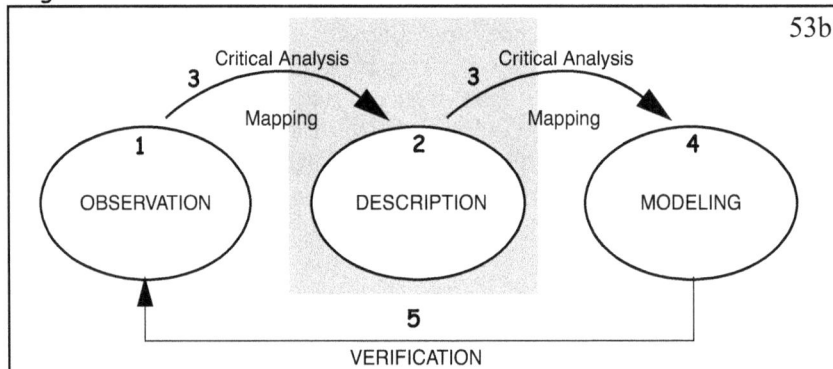

53b

In the background our brain is performing pattern feature extraction and building internal brain models simultaneously and continuously. Often we may not be aware or conscious of this process. When the process and its models are brought to mind and subjected to analysis, a critical thought process is begun.

**Step 3. Critical Analysis**  Mapping the described features onto a model system requires a critical analysis (shaded in gray on FIGURE 53c) of the observations.

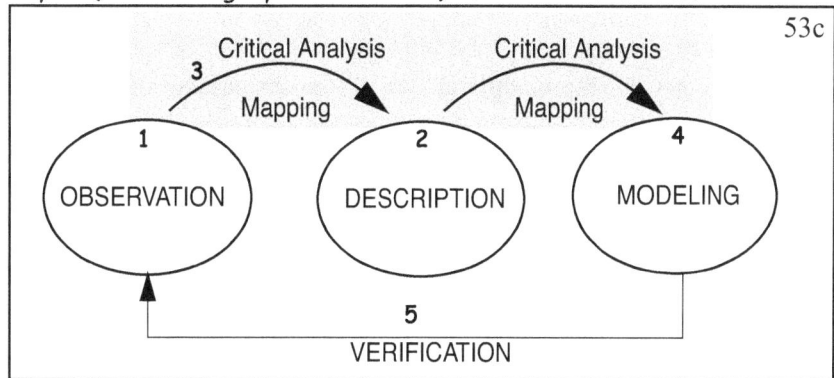

- The **observables** of the system describe what properties can be measured and how a set of properties define a state or condition of a system. The particular values of those properties allow categorization of the system.

- How the **elements** will be represented in the model must be determined. For example, will words, latin phrases, musical notes, chemical notation, mathematical symbols, or colors be used to symbolize each element in the system?

- The characteristics of the **background space** in which the mapping will be formed is necessary. Usually the background space has the same number of dimensions but may be scaled larger or smaller. However, as the pictured globes show, this does not have to be the case.

- A definition of the **rules** that arrange the elements of the structure being described into the background space is required. Elements can be ordered:

    - By naming and counting (nominal ordering) - who and how much there is.
    - By ordinal measure (ordinal ordering) - arranging the elements by height, weight, or in the order that timed events happen.

  • <u>By arrangement in space (spatial ordering)</u>, e.g., 30° north latitude and 15° west longitude.

Critical analysis is the aspect of knowledge construction that depends on logic, rational analysis, precision in language, principles of mathematical thinking (including sorting, ordering, categorizing, transforming), and discovering ideas of number, form, and difference. The skills of critical analysis are the mapping skills used to build a model of some aspect of the world.

**Step 4.**
**Modeling**
Once the mapping process is completed, a model exists (shaded in gray on FIGURE 53d). A model is a representation of the world that exists only as an abstraction. It is created by the process of pattern feature extraction and critical analysis.

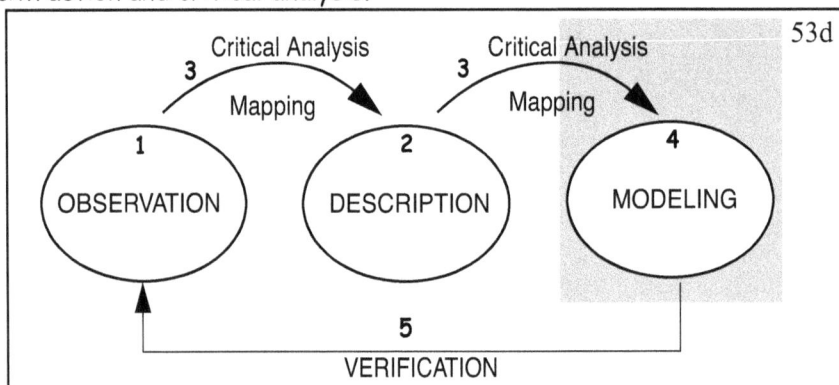

**Step 5.**
**Verification**
The methods of verification and error checking (shaded in gray on FIGURE 53e) in a critical mode of thinking require a willingness to define and formally represent an idea or observation, communication, and acceptance of another's critical analysis of an idea.

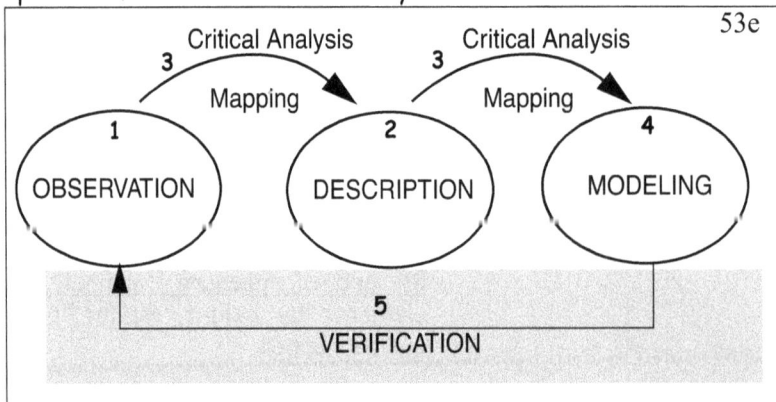

This approach to modeling works for descriptive, explanatory or experimental models and is used recurrently in the Progression of Inquiry (detailed in the next chapter).

**FIGURE 54.** The Progression of Inquiry.

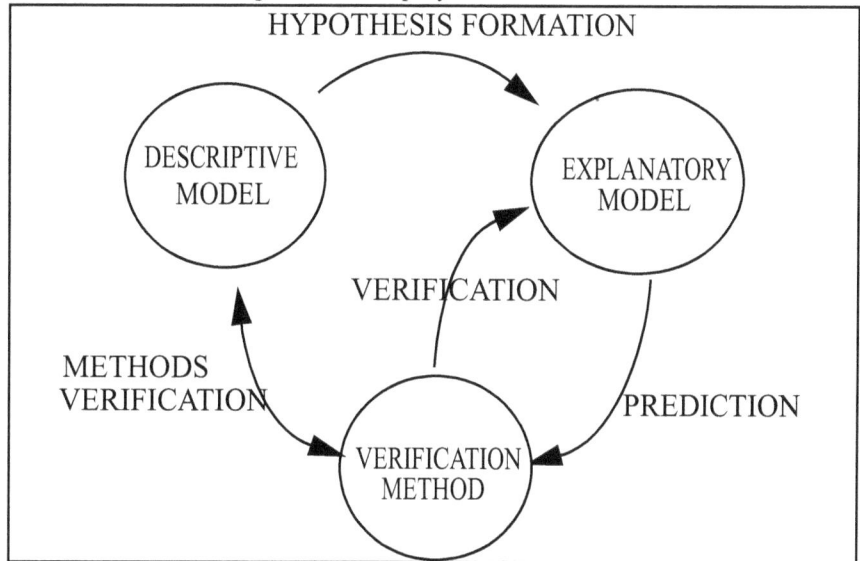

The brain obtains information from its observations of the natural world and from its knowledge of the formal subjects that it has learned.

All of this information is mapped in the human brain onto knowledge surfaces that possess cusps. Therefore, all forms of human-modeled knowledge are subject to the strengths and weaknesses inherent in the knowledge surfaces that we have studied earlier.

# CHAPTER 12

# The Progression of Inquiry is the Scientific Method

## The Scientific Method

Science is a particular way to look at and explore the world. It is a habit of mind for building knowledge. Modern civilization is built on the idea that the best method for advancing knowledge is through the human endeavor of experimental and critical science. Modern science has evolved over the last four centuries to become the highest form of critical exploration with its careful methods of verification and the evaluation of ideas based on data and measurement.

The scientific method is an organized way of exploring and describing the world and proceeds in a series of steps. The characteristics of the scientific method are:

· Observation.

· Making models and proposing hypotheses.

· Testing and retesting by experimentation.

There are two sub-divisions to the scientific process, one **theoretical** and the other **experimental**.

· In theoretical methods, observations about the world are organized into a model of the world.

· In experimental methods, the theoretical model is tested by making observations of an experimental model.

· In the scientific method, the theoretical model is compared to the experimental model.

- 👍 If they agree, the experiment supports the theoretical model.

- 👎 If they disagree, the experiment rejects the theoretical model.

## Science Begins With Observation

All science begins with observations of the patterns of the natural world. A knowledge of the patterns of natural events and the resulting effects are among the most basic interests of all people. Since the dawn of our species, survival has depended upon recognizing these patterns which have included day-night cycles, the motions of the sun and moon, the turn of the seasons, the migration of animals, the rhythm of human existence.

Archeological evidence from cave paintings and the notching of bone and reindeer horns suggests that pre-historic humans were extremely careful in their recording of seasonal and temporal patterns. Such knowledge is acquired by simple observation. However, observation alone is not "modern" science. Observation is "proto-science". Proto-science accepts observations without question or verification.

## Skepticism in Modern Science

In modern science, what is known is viewed skeptically. A **skeptical** view is characterized by the continual questioning of the certainty with which something is considered "true". "How sure am I, that I actually know what I think I know" is the skeptical view. This includes a skepticism of the ability of the human to objectively make observations. The scientific methods of error-checking use the experimental examination of a natural system to verify what is known. Modern scientific thinking requires proof of how cause and effect are connected and also that correlations are not just due to chance.

## Modeling is the Core of the Scientific Method

Relating a cause to an effect is the purpose of a model. In a model, a particular observable (a cause) is linked to another particular observable (the effect). The cause is an observable that an experimenter will vary. The effect is an observable that the experimenter has theorized to be related to the cause. A working model explains how things happen. Investigating cause and effect uses logic, inference, and deduction.

In modern science, the fundamental rules of the model are discovered from careful consideration, measurement, experiment, and analysis of simple cases. Observation is the first step in constructing a descriptive model. The measured attributes first are mapped onto a descriptive model then later a theoretical model. The models are usually made from a "formal system" of equations, words, or geometric figures.

The validity of the model is tested by:

· Making a prediction based on the hypothesis.

· Performing experiments in the natural world to test the model.

· Making experimental measurements.

· Recognizing that the observer may influence the experiment and measuring that influence.

· Changing the model when the experimental evidence requires modification.

## Modern Science is Based on Evidence

The explanatory cause-effect model is constantly modified with experimental evidence that verifies or falsifies the ideas on which the model is constructed. In experimental science, the relationship between the cause-observable and the effect-observable is called a **hypothesis**.

· A hypothesis is not just a question - it is a specific kind of question.

· A hypothesis states: **The cause is related to the effect in the following way. Is this relationship true or false?**

## Verification in Modern Science

In order to verify with some certainty the validity of a hypothesis, experimental-skeptical methods include:

· **Control** experiments to be certain that the effect observed is related to the proposed cause.

- **Calibration** of measurements to be sure that an agreed upon standard is used to quantitatively compare measured effects between experiments and experimenters.

- **Context** of the system under study with reference to other systems.

- **Communication** between experimenters to confirm and evaluate each other's experimental and theoretical work.

This overall process is the Modern Scientific Method expressed in the complete Progression of Inquiry.

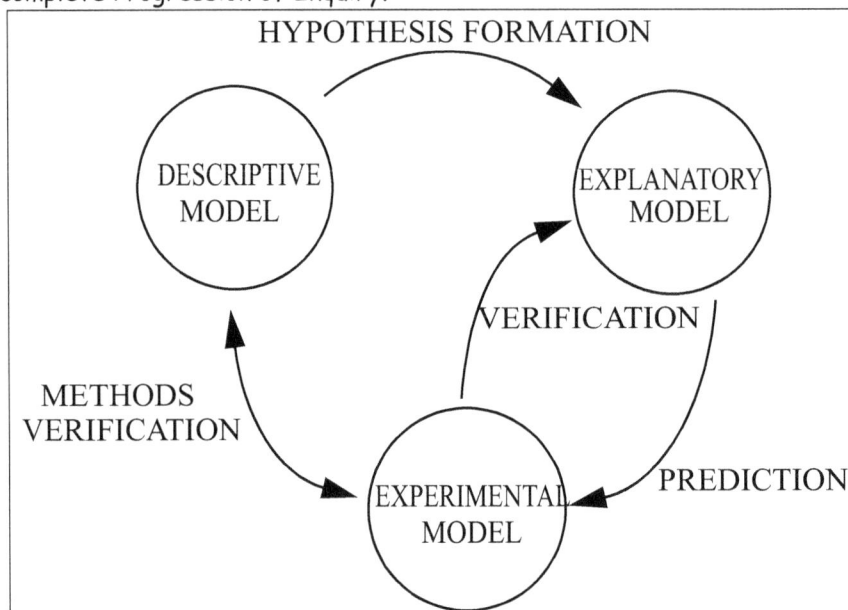

**FIGURE 55.** The Progression of Inquiry is the Scientific Method and uses sequential model making.

## The Scientific Method

In order to emphasize the process, the steps used in the practice of science are now reviewed. The following description includes more detail on the two subdivisions of the Scientific Method, theoretical and experimental branches.

### Theoretical Steps

1. **Observations** are made about some part of our world. For example, we may notice that the sky is blue. The observation is described by a Language of Pattern analysis. Scientific observation accounts for and measures:

- The properties of the systems,
- The elements or parts of the system,
- The way the elements are arranged,
- The background space where the elements are arranged, and
- The background conditions or properties under which a particular arrangement is found.

2.   A **descriptive model** of the observations is made. Symbols represent the observed and measured elements. Examples of symbols are:

- words,
- mathematical expressions,
- musical notes and writing,
- dance steps,
- cultural rituals,
- electronic schematics, and
- chemical element notations.

3.   **Cause and effect** create explanatory models. Reasoned hypotheses are made of the relationships of the system. These rules tell us how one element or rule (a cause) will affect a systems property (the effect). A model that explains how system properties emerge from the relationship of elements, rules of arrangement, and the background space is called a **theoretical** model.

**Experimental Steps**   In the scientific method, the theoretical model must always be tested by **experiment**. **Evidence** in the scientific method comes from careful **experimental design**. The method of experimental science is described as a series of steps.

1.   **Define the problem to be explored.**
- Usually a specific question is asked that focuses attention on one aspect of a model. This focus is often labelled the **problem** that is under investigation.
- For example:   a scientist might be interested in the problem: "Where do animals get oxygen?"

2.   Make a hypothesis.

- A **hypothesis** is a special kind of question. It proposes a relationship between two parts of a system.
- The relationship can be between elements, elements and the background space, the rules of arrangement and the elements, or even different rules.
- A good hypothesis proposes, "If I make this change in this part, I will be able to observe that effect on that part".
- The relationship proposed by the hypothesis leads the experimenter to make a prediction. A **prediction** is the logical outcome of the hypothesis and how the model works.
- A good hypothesis offers a single proposition that can be shown to be true or false by experiment.
- With a good hypothesis, an experiment can be designed that specifically tests to see if the proposed relationship actually exists.
- A conclusion is reached that the proposed relationship exists if the prediction matches the actual observed result from the experiment.

3.  Design an experimental system to test the hypothesis.
    - An experimental system is a working model of the real world. However, it is a simplified or **controlled environment**.
        - A controlled system is one in which every aspect of the system can be defined and described. Therefore, it can be repeated by someone else at another place and time.
    - Experiments generate data to explore how one property in the system (the **cause**) is related to a second property (the **effect**). Experimenters obtain this data by making and recording measurements.
    - Both of these properties, the cause and the effect, must be able to be observed and measured. They are special experimental properties called **observables**.
    - In an experiment, the measured values of observables can change. Observables that change or vary their value are usually called **variables**.
    - If the experimenter varies or changes the value of one of the observables, this is called the **independent variable** (the cause).

- The variable whose value is changed by the independent variable is called the **dependent variable** (the effect).

4. Gather data by measurement and observation from the experimental system.
- Data are collected by measuring and observing the variables in the experiment.
- The relationship between the cause and effect properties is determined by analyzing the data.
- Experimental results are compared to a **control experiment**.
  - Scientists are skeptical of their own experiments, therefore every experimental system has a **control experiment**.
  - A control experiment is an experimental set-up that is used as a **standard** or **reference** for the system.
- This means that everything in the control and experimental set-up are the same except that the independent variable (the cause) is taken out of the control experiment.
- If the dependent variable (the effect) changes in the control experiment, the changes in the experimental system can not be attributed to the independent variable (the "cause" was never there!).
- Importantly, experimental systems all contain degrees of uncertainty: uncertainty in measurement, variability in the natural response of the system to produce an observable measurement. These must be taken into account.

5. Draw a conclusion that supports or rejects the hypothesis.
- If the experimental data gives the same result as that predicted by the hypothesis within the degree of uncertainty, then a conclusion can be made that the hypothesis is supported.

6. Repeat the experiment; have other scientists repeat it.
- Since the scientific method is skeptical, the experiment is done many times by many different scientists. If the same conclusion is reached by everyone, then the hypothesis is generally accepted as a fact by the scientific community.

- As more and more of the ideas that went into building the model are accepted as facts, the model may be accepted as a **scientific theory**.

- A **scientific law** is different from a scientific theory. If after many, many tests by many, many scientists in many different situations, the relationship proposed by a hypothesis is always proved correct, it may be called a scientific law.

As a result of the success of the scientific method in exploring and understanding the natural world, the scientific model of investigation has become the *de facto* standard for virtually all forms of critical human inquiry. This includes fields of art, language, and social science.

# CHAPTER 13

# Getting Started with the Language of Patterns

## The Language of Patterns Vocabulary

The Language of Patterns is extremely useful throughout the process of knowledge construction and critical thinking. It enables the early use of modeling, which is the key tool that promotes effective problem solving.

It is a simple method that enables critical thinking.

The Language of Patterns has six major ideas, which work together.

These are:

· Systems

· Properties

· Elements

· Rules

· Background Space (or Context)

· Evolution or dynamics (Change over time)

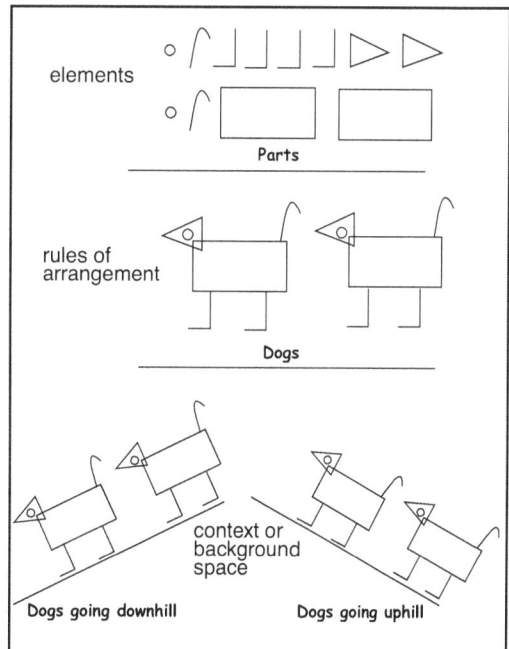

**FIGURE 56.** A Language of Patterns description of a dog

The Language of Patterns is used to describe the construction of information and knowledge in terms of pattern features. These pattern features are the elements, rules, and background space of an object, idea, or system that give rise to system properties.

- **Systems** are characterized by **properties** (color, shape, texture, size, weight, states).

- System properties emerge from the interaction of system elements governed by certain **rules** in a **background** or **context space**.

**Systems** result when elements are put together.

- Every system is composed of a certain group of parts or events that we will generally call **elements**.

- In a system, the elements are arranged somewhere - in what is called a **background space**.

- The **arrangement** of elements is governed by **rules of relationship** between each element **and** between the elements and the background space.

This is an example of a system:

- The **elements** of this system are seven objects each of whom is a *subsystem* whose **properties** are that they are squares with varied fill patterns.

- The **background space** in which they are arranged is a flat page [in technical terms: a plane of two dimensions (length and width)].

- One of the **rules** that arranges the squares is that their center-points lie in a straight line.

## Properties and Observables

We can describe each system in terms of its own **properties**. A property is a characteristic of a system (structure, object, event) that describes it. By comparing and contrasting the properties or attributes of different systems, we can group and sort them. These are fundamental critical analysis skills. It is important to recognize that the elements that compose a system are themselves subsystems and have properties that also allow them (the subsystem) to be grouped or sorted.

Properties form the basis for sorting, grouping, comparing, contrasting, and categorizing things.

- To fully describe a system, every property would have to be listed and enumerated.

- This is usually impossible and unnecessary since most systems have many, many properties. We hardly ever have to know or name every property in order to be able to usefully sort, contrast, or compare systems.

When we work with systems, we choose a property that we can easily observe or measure. The particular properties that are chosen to observe or measure are called **observables**.

- An important part of critical thinking and analysis is to be able to specifically state what the observables are that are being used to characterize a system.

- The choice of an observable is almost always an assumption. Careful skeptical critical inquiry must always question that choice.

- **CAUTION**: The choice of observables can turn out to be the limit, prejudice, or error in the analysis of a system.

The following example demonstrates the advantage of picking the correct observable. Consider this system of objects:

**FIGURE 57.** The property o number is useful seat this system of people.

If we need to know how many chairs to set up in a classroom so that everyone can have a seat, the proper observable would be the number of people. Having the properties of their names, favorite color, food, baseball team, religion, and income might be interesting, but it would not help us set up the room. The problem of picking the wrong observable can also be demonstrated by using the same system of objects:

**FIGURE 58.** Citizenship and registration to voter is useful observable for the same system of people if they are to vote.

This group of people is going to vote for President of the United States. The correct observable is whether they are citizens of the USA and are registered to vote. If we tested them for the ability to read and used that to sort them into voting and non-voting groups, in 2012 this would be against the law. [The Civil Rights Act of 1964 made it illegal to use a test of literacy as the observable to decide whether a citizen of the US could vote or not.]

## Analyzing Systems

When we want to describe a system, we need to describe the elements in it, the background space where the elements are arranged, the rules that arrange the elements in the space, and the properties that emerge from all of these system interactions.

When we look at our world, it is the systems that we notice. Cars are systems, people are systems, poems are systems, songs are systems. The package of elements, rules of arrangement, and background space gives rise to the characteristics (observable properties) that make us laugh at jokes, cry at tragedies, fall in love, see the stars twinkle, and navigate by the constellations.

The Language of Patterns gives us the critical analysis tools to explore systems and their properties, yet it reaches its full power when turned to the exploration of the most compelling and interesting aspects of our existence, how the world **changes** and **evolves**. These dynamic processes occur when structures or systems are moved to action. It is the dynamic analysis of change and evolution for which the Language of Patterns was designed. We will learn how this important task is accomplished in the final unit.

# CHAPTER 14

# The Progression of Inquiry and SymmetryScience™

## Systems Modeling and Teaching Science

**Weaving the Strands of Learning**

With the vast quantity of information available today, the greatest challenge in education is to harmonize these three strands:

- How the brain processes information and yields knowledge

- How the endeavor of gathering information has led to a knowledge of our universe

- How to structure a coherent learning program for all citizens, especially our children

The Language of Patterns and the Progression of Inquiry underlie a system for learning science, The SymmetryScience curriculum. Everyone who practices scientific exploration through the Progression of Inquiry benefits from the system modeling tools found in this book. It is through the Language of Patterns and the Graphical Analyzer System that these tools can be applied to scientific inquiry.

This book can be used to help prepare teachers, parents and science professionals involved in science education to teach the SymmetryScience™ program. In this chapter the SymmetryScience™ path will be briefly explained as an example of using systems modeling generally in science education.

The SymmetryScience™ curriculum teaches and uses the Language of Patterns as the critical thinking language in which scientific knowledge and inquiry can be learned and expressed. The Language of Patterns is learned through the Graphical

Analyzer System. The Graphical Analyzers are a practical series of knowledge organizers that guide the student to progressively learn science inquiry while teaching the Language of Patterns.

Mastering anything of importance requires time, work, and inspiration; progress is made by the continuous and diligent application of all three. Knowledge comes from understanding, which relies upon information. The scientific method of inquiry is the tool that must be mastered to move freely from information to understanding. As each person grows in education and develops in mind, there is process guided by the Cycle of Pedagogy that hones this tool of understanding and knowledge. The Language of Patterns, through the Graphical Analyzer System, is a map on this "road of inquiry".

## The Graphical Analyzer System

The following chapters will introduce and guide your preparation for using the Graphical Analyzer System.

In practice, the Graphical Analyzers are intended to structure how all students, from their earliest stages of learning to advanced study in science, use observation, description, measurement, modeling, and experimental design to learn and practice science inquiry.

- Using the Analyzers continuously and progressively throughout a student's schooling (from kindergarten to high school), the SymmetryScience curriculum guides the learner to practice the common threads of scientific inquiry and knowledge construction. With practice and structure, the students will achieve a core competence and natural fluency in using the scientific method to explore and understand the world.

- Each Analyzer is introduced and demonstrated in the final Unit. Examples showing its application to learning science and critical thinking are linked to the appropriate grade levels.

**Interdisciplinary thinking** is vital to the SymmetryScience program. The SymmetryScience method encourages the use of the Graphical Analyzers across the curriculum in order to emphasize and widen a student's opportunity to practice the scientific method and to understand the general value of the method of scientific and critical inquiry in all fields of interest and all walks of life.

- In this way, a student will be guided to make observations, descriptions, measurements, model the world, and express themselves using evidence in their studies in literature, history, geography, mathematics, and the arts.

- This important aspect of the SymmetryScience approach encourages the students to use the scientific method constantly in their daily life, making scientific and critical inquiry a habit that extends to all areas of their interest.

- Examples of this cross-curricular process are also given in this final Unit.

<table>
<tr><td>

**The Organization of the Graphic Analyzer System**

</td><td>

The Graphic Analyzer System is organized around the vocabulary of the Language of Patterns. There is a variety of related graphic analyzers that support the basic ideas used in the Language of Patterns:

- Systems
- Properties
- Elements
- Rules
- Background
- Change

As recalled in the diagram below, the Language of Patterns is closely linked to these steps in the scientific and critical thinking process and are guided by the Progression of Inquiry's successive models.

</td></tr>
</table>

1. Observation

2. Descriptive Modeling

3. Critical Analysis and Hypothesis Mapping

4. Explanatory Modeling

5. Verification by prediction and testing in an experimental model.

6. Experimental Modeling and Experimental action.

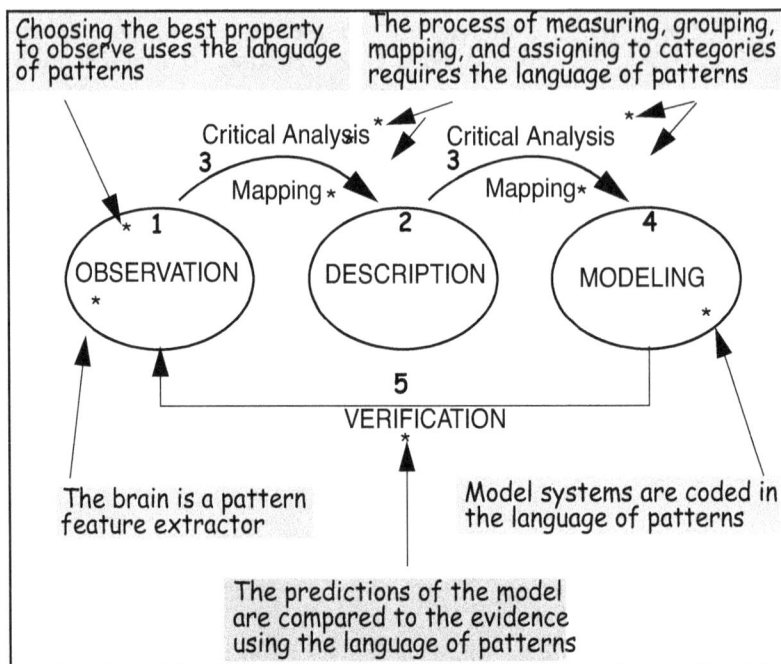

**FIGURE 59.** Critical analysisis codified by the language of patterns and graphical analyzers.

The following chart indicates the Graphical Analyzers which are used with the five basic concepts of the Language of Patterns. The critical thinking skills and the scientific inquiry process associated with each Graphical Analyzer is listed.

| Language of Patterns Concept | Critical Thinking Skill | Scientific Inquiry Process | Graphical Analyzer |
|---|---|---|---|
| Systems<br><br>Elements | Properties<br>Observables | Observation<br>Description | Element Analyzer<br>Element Counter<br>Element Extractor<br>Property Analyzer |
| | Sorting<br>Classifying | Observation<br>Description<br>Mapping | Sorting Analyzer (single observable)<br><br>Sorting Analyzer (multiple observable)<br><br>Sub-sorting Analyzer<br><br>Classifying Analyzer<br><br>Compare and Contrast Analyzer<br><br>Conflict Analyzer |
| | Describing<br>Mapping | Description<br>Mapping<br>Modeling | Technical Writing Analyzer<br>Coding Grid |
| Background Space | Properties | Observation<br>Description<br>Modeling | Background Space Analyzer<br>Boundary Analyzer |
| | Measuring<br>Mapping | Description<br>Modeling<br>Verification | Background Feature Extractor |

| Language of Patterns Concept | Critical Thinking Skill | Scientific Inquiry Process | Graphical Analyzer |
|---|---|---|---|
| Rules | Arrangements | Observation<br>Description<br>Mapping | Arrangement Analyzer |
|  | Inferring rules of arrangement | Modeling<br>Verification | Rule Extractor |
| Structure | Describing | Observation<br>Description<br>Mapping<br>Modeling | Structure Analyzer<br>Modified Structure Analyzer |
| Evolution or Change | Describing | Observation<br>Description<br>Modeling | Evolution Analyzer |
|  | Predicting | Modeling<br>Verification | Evolution Analyzer |

The next grid shows the progression of scientific inquiry through each grade level in SymmetryScience™.

| SYMMETRYSCIENCE AND THE PROGRESSION OF INQUIRY | | |
| --- | --- | --- |
| TYPICAL GRADE LEVEL | SCIENTIFIC/CRITICAL THINKING | STEP OF SCIENTIFIC INQUIRY |
| Kindergarten | Observation | Exploring the parts or elements of a system, the properties of these elements, and the arrangements of system elements into patterns. Increasing ability to pay attention - set maintenance. |
| Grade 1 | Description | Describing and defining system elements, properties, arrangements, and background space. Classifying by characteristics. |
| Grade 2 | Measuring and Description | Measuring elements, properties, arrangements, and background space and learning about cycles and processes of change. |
| Grade 3 | Descriptive Models / Cause and Effect | Developing skepticism of observation and considering cause and effect. |
| Grade 4 | Inferring Observables into Descriptive Models | Beginning to draw general conclusions from diverse information. Emergence of abstract thinking. |
| Grade 5 | Modeling and Verification | Proposing simple linkages of cause and effect. Increased ability in set-shifting/cognitive flexibility. |
| Grade 6 | Testing Explanatory Models with Verification | Constructing models of cause and effect. Forming hypotheses and testing for verification. |
| Grade 7 | Experimental Models and Design | Testing a hypothesis of a proposed model of cause and effect by experiment. Designing and using controls and variables to make conclusions. |
| Grade 8 and Above | Experimental Models and Design | Inferring unobserved properties from observed properties (observables). |

## Using the Graphical Analyzer System

We are now ready to examine the specific Graphical Analyzers in the System. The Graphical Analyzers are described along with examples to demonstrate each Analyzer. Practice Exercises will include Analyzers that show how science content can be organized. Occasionally, the use of an Analyzer across the curriculum will be demonstrated by giving examples outside the science content areas.

# Unit Five

# The Graphical Analyzers

# CHAPTER 15

# Structure and System Analyzers

## Picturing and Describing a System

| Analyzers | Application |
|---|---|
| System Analyzer

Structure Analyzer | Characterize the composition of a system or structure in terms of its properties, elements, rules, and background space |

**Exploring the World with the Language of Patterns**

The world around us is formed as structures and arrangements of structures into systems. We can divide the kinds of arrangements in the world into two major categories: those that are obvious on observation and those that are not. This is an important example of the progression of inquiry. Critical scientific analysis is used to understand both types.

**Observed systems**

These are systems in which the elements or parts are arranged in a fashion that is obvious on observation. When we look at these systems, we try to understand how they are assembled and how they work. Typical examples are:

· Fields of naturalist study, such as:
  · Patterns of day and night,
  · Patterns in clouds and weather,
  · Patterns and behaviors of biological organisms,

- Patterns of geological formation.
- Technical structures, such as:
    - Parts and schematics of machinery and tools,
    - Organizational charts of social groups (businesses, governments, law).
- Structures in artistic works, such as:
    - Chapters in novels,
    - Scenes and acts in dramatic works,
    - Stanzas in poetry,
    - Verses and refrains in songs,
    - Movements in instrumental works.

**Inferred systems**   In many systems, a group of objects are linked in a fashion that is not so obvious with simple observation. The connections between the elements must often be inferred or deduced. Examples of these include:

- The classification of the chemical elements based on observable properties into the Periodic Table. The organization of the Periodic Table infers the internal, unobservable structure of atoms.

- The classification of living things into a systematic organization. This organization is necessary for the process of evolution to be inferred and understood.

- Developing an understanding of the motion and evolution of stars and planets from observations of the light and motions of the planets and stars in the sky. These explorations allow understanding of the birth and evolution of the expanding universe by inference.

- Understanding processes outside of the realm of "natural sciences". Examples include:
    - The hidden relationships between characters in a story or work of art.
    - The explorations of shape and form and how they relate to natural objects in the art of Picasso, Monet, or Leonardo da Vinci.
    - Bach's exploration of the ordering of notes in the fugue.

The observed systems can be explored successfully with observation, while inferred systems require the tools of logic and inference in addition to observation. Both require that we can explore a system or systems of objects and describe the elements, rules, and background space. Thus, all critical understanding starts with a careful Language of Patterns analysis.

## The Graphical Analyzer System in Scientific Inquiry

Scientific investigation is concerned with inquiry into how a system is arranged. Sorting and grouping are natural tasks for the brain; in scientific investigations, they are the starting point for deeper inquiry into a system and its properties. This task is approached through a more thorough analysis of the properties and arrangement of the elements in the system.

The role of the Graphical Analyzer System is to guide the description of a system or structure for critical analysis. To appreciate the value of such a method, try this apparently simple (but deceivingly difficult) task:

In words, write a description of this system (structure) in the box below. You will be surprised how hard it is to communicate the information clearly and completely.

Frustration is commonly felt from not knowing how to begin this task. This is the same problem shared by teachers and students at every grade level when confronted by word problems, essay questions, and other forms of open-ended questions.

· Read your description to a friend and have them draw it from your description.

· Compare your friend's drawing to the actual system.

· You will likely confirm the difficulty that you experienced in the description and communication of such an apparently simple system.

## System Analyzer

Any system or structure or sub-system (element) can be described using a System Analyzer, which is comprised of a Structure Analyzer and a Property Analyzer.

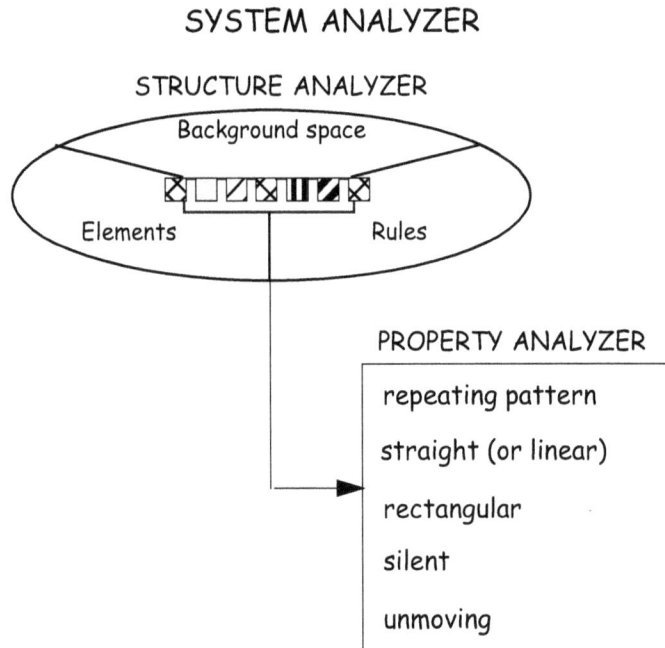

SYSTEM ANALYZER

STRUCTURE ANALYZER

Background space

Elements                    Rules

PROPERTY ANALYZER

repeating pattern

straight (or linear)

rectangular

silent

unmoving

The System Analyzer has many of the characteristics of a simple concept map or "brain-storming" diagram currently in common use in many classrooms and work environments. The structure analyzer will be discussed below. The Property Analyzer will be discussed in the next chapter.

## Structure Analyzer

The Graphical Analyzer System provides a methodical process to accurately describe and communicate that is widely required in science. The Structure Analyzer provides an overview of what we need to discover about a system in order to describe it scientifically. A language-of-patterns analysis forms the basis of describing structures:

- Every structure is comprised of a certain group of **elements.**

- In a structure, the elements are arranged somewhere - in what is called a **background space**.

- The **arrangement** of elements is governed by **rules of relationship** between each element and between the elements and the background space.

- The Structure Analyzer is differentiated from the system analyzer because the property analyzer has been removed. This emphasizes: **properties are emergent** from the interactions of elements, rules and background space.

This concept of structure is represented in the Graphical Analyzer System with a Structure Analyzer.

The Structure Analyzer can be used to quickly see the overall arrangement of an object or system of interest. A quick structure analysis often leads to a knowledge of what is needed to complete the description of the system.

This analyzer reminds us that structures are composed of a set of **elements** arranged by certain **rules** in a **background space** that will give rise to certain properties. The Structure Analyzer requires that its elements, rules, and background be defined and described.

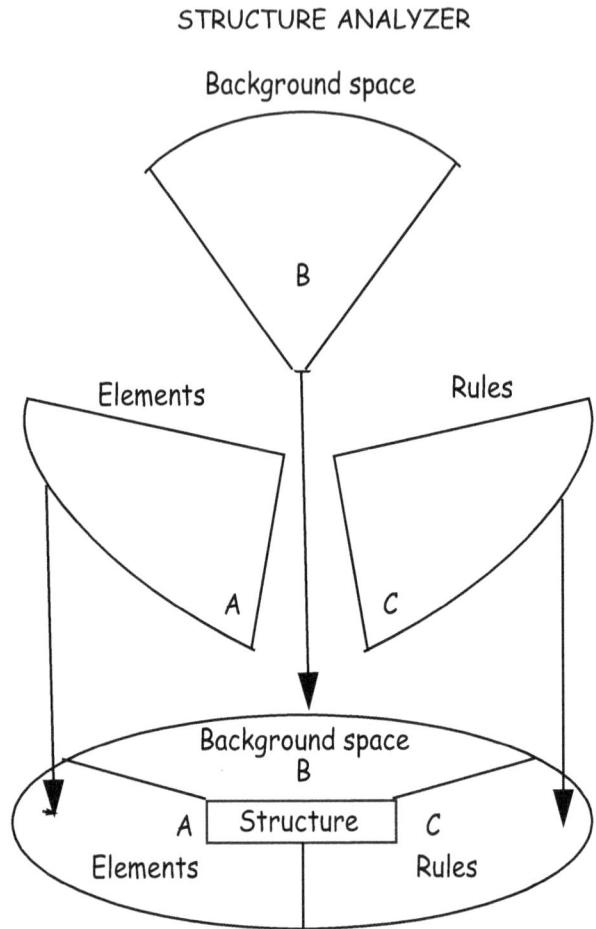

STRUCTURE ANALYZER

Background space

B

Elements

Rules

A

C

Background space
B

A    Structure    C

Elements         Rules

In the pattern-filled squares structure example:

- The **elements** of this structure are the seven pattern-filled squares.

- The **background space** in which they are arranged is on a page of two dimensions (length and width).

- One of the **rules** that arranges the squares is that they are ordered in a straight line.

STRUCTURE ANALYZER

Background space

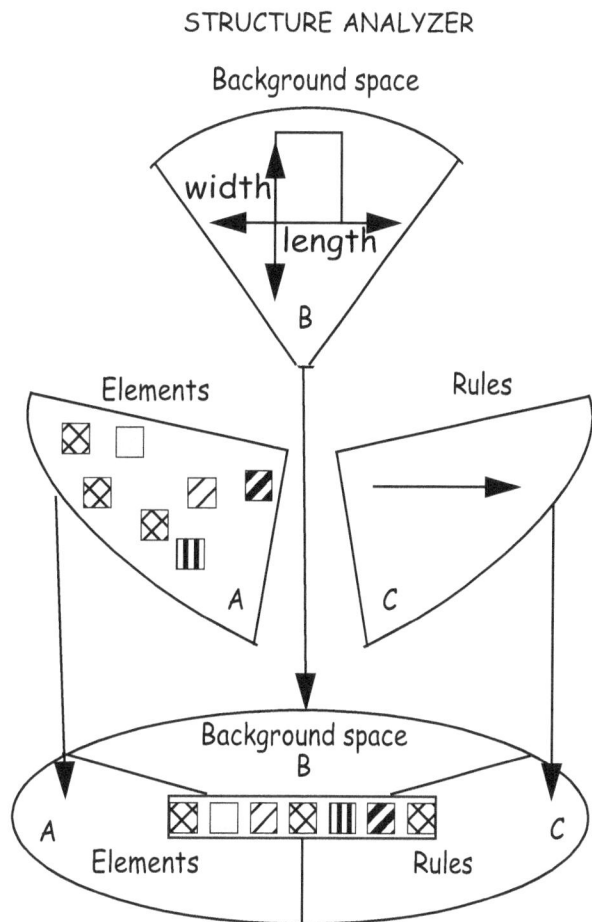

To fully describe the structure, rules to order the squares and to describe the internal fill pattern must also be included. This will be set forth in the coming chapters. In the next chapters, we will also explore the Graphical Analyzers that will help guide the discovery of elements, background space, and rules. Finally, we will see how to describe a system or structure when it changes or evolves.

# CHAPTER 16

# The Element Analyzer Series

## Properties, Observables, and Elements

| Analyzers | Application |
|---|---|
| Property Analyzer | Identify and characterize properties of objects and systems |
| Element Analyzer | |
| Element Counter | |
| Element Extractor | |
| Technical Writing Analyzer | Systematically describe a pattern, structure, or system |
| Coding Grid | Exchange objects and symbolic forms to build codes |

Systems have **properties** that result from the arrangements of the elements in their background space. These properties are what give a system of structure its qualities that let us sort, group, and categorize it. Properties are not what makes up the structure or the system; they are the overall emergent <u>qualities</u> of the system. The properties that we measure and observe that describe the **state** of a system are called **observables**.

## The System Analyzer

For reference, we repeat this graphic analyzer. A System Analyzer fully describes any system by adding the property analyzer to the structure analyzer.

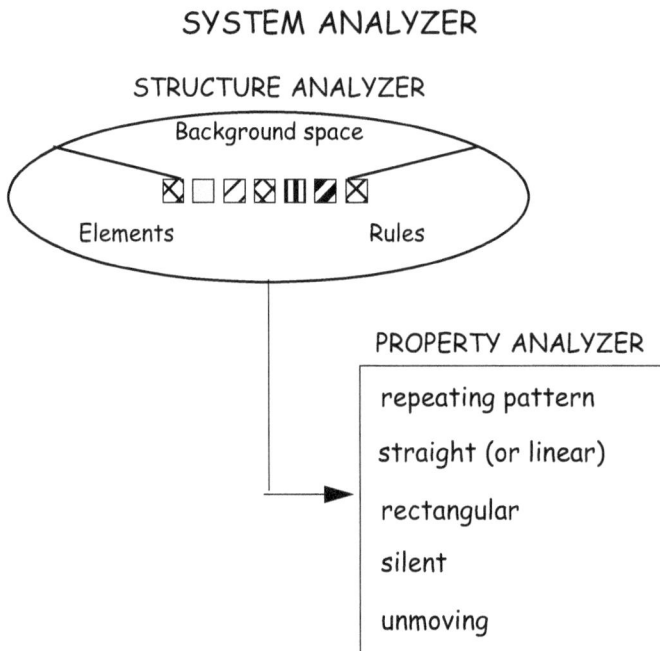

### SYSTEM ANALYZER

STRUCTURE ANALYZER

Background space

Elements          Rules

PROPERTY ANALYZER

repeating pattern

straight (or linear)

rectangular

silent

unmoving

Any system or its parts [properties, elements, rules, background, changes] can be described using a **Property Analyzer**. In addition, it is can be useful to analyze a property as a system itself. This will focus analysis on how the emergent-property can be characterized and therefore measured [by choosing useful observables].

# Property Analyzer

Some properties for the pattern system example have been entered on a Property Analyzer. Can you think of some others?

On the following page, a variety of named systems have been placed in the header section of a Property Analyzer. List some properties of these systems. Properties are attributes and qualities of a system such as its color, texture, degree of difficulty, etc. Some properties have been filled in to get you started. These Analyzers may be arranged so that the properties used to describe a system become new systems that themselves can be analyzed. This process is also expressed in the more familiar, but somewhat less formal "concept map".

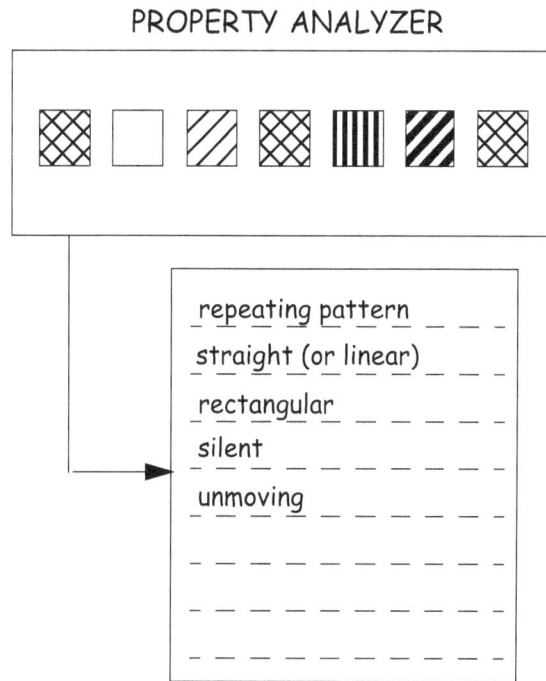

PROPERTY ANALYZER

repeating pattern _ _ _ _

straight (or linear) _ _ _

rectangular _ _ _ _ _

silent _ _ _ _ _ _ _

unmoving _ _ _ _ _ _

_ _ _ _ _ _ _ _

_ _ _ _ _ _ _ _

_ _ _ _ _ _ _ _

**Practice Exercise** Complete the following exercises using the Property Analyzer. (Possible answers appear on the next page.)

## PROPERTY ANALYZER

Mammals

warm-blooded
produce milk for young
furry or hairy

## PROPERTY ANALYZER

Water

## PROPERTY ANALYZER

Liquids

## PROPERTY ANALYZER

Thunderhead Clouds

PROPERTY ANALYZER

PROPERTY ANALYZER

**Answer** Possible answers to the exercises on the previous page.

Mammals

warm-blooded _ _ _ _ _
produce milk for young _
furry or hairy _ _ _ _
four chambered hearts _

_ _ _ _ _ _ _ _

_ _ _ _ _ _ _ _

_ _ _ _ _ _ _ _

_ _ _ _ _ _ _ _

Water

wet _ _ _ _ _ _ _ _
can freeze into ice _ _ _
evaporates _ _ _ _ _ _
has weight _ _ _ _ _ _
is a liquid _ _ _ _ _ _
fun to play in _ _ _ _ _

_ _ _ _ _ _ _ _

_ _ _ _ _ _ _ _

Liquids

takes shape of container _
flows _ _ _ _ _ _ _ _
can become solid if cooled _
can become gas if heated _
has weight _ _ _ _ _

_ _ _ _ _ _ _ _

_ _ _ _ _ _ _ _

_ _ _ _ _ _ _ _

Thunderhead Clouds

float in air _ _ _ _ _ _
lightning can jump to Earth
dark _ _ _ _ _ _ _ _
high and towering _ _ _
heavy driving rain _ _ _
can make hail _ _ _ _ _

_ _ _ _ _ _ _ _

_ _ _ _ _ _ _ _

## Cross-Curricular Example

Here are several examples using the Property Analyzer across topic domains.

PROPERTY ANALYZER

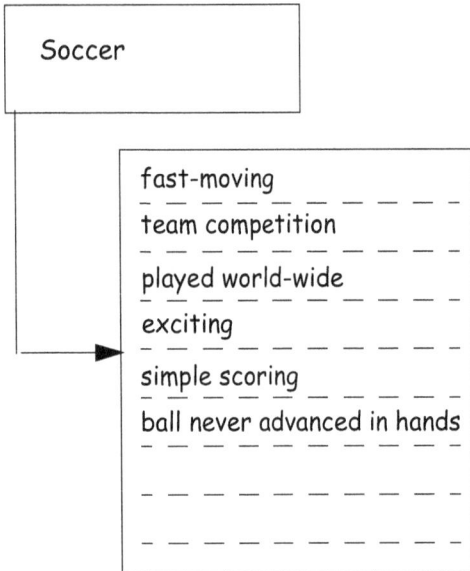

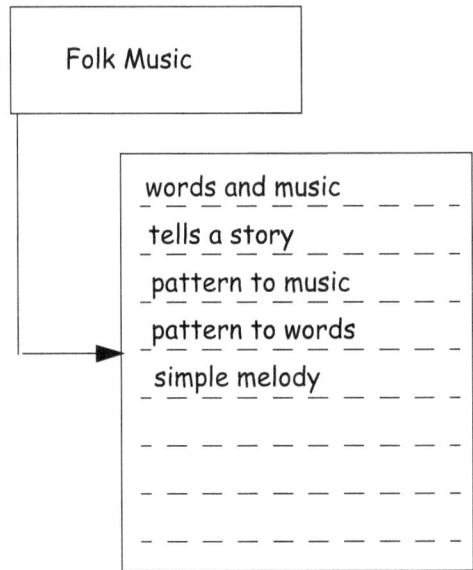

Soccer

fast-moving

team competition

played world-wide

exciting

simple scoring

ball never advanced in hands

PROPERTY ANALYZER

Folk Music

words and music

tells a story

pattern to music

pattern to words

simple melody

PROPERTY ANALYZER

## Element Analyzer

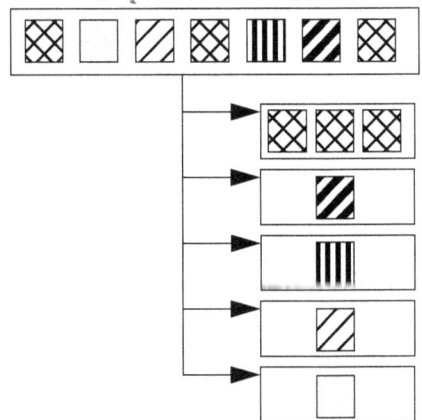

We can explore the example of the previous geometric pattern using the Element Analyzer.

If the elements in the pattern example are sorted by the **observable** of their fill pattern, the system of squares can be broken down. This manipulation separates the elements of the system into its individual parts. The categorization of the elements by observables suggests that they are "sub-systems".

ELEMENT ANALYZER

Identifying the elements of an observation is an important scientific activity. A first grade sky watch might generate the following Element Analyzer.

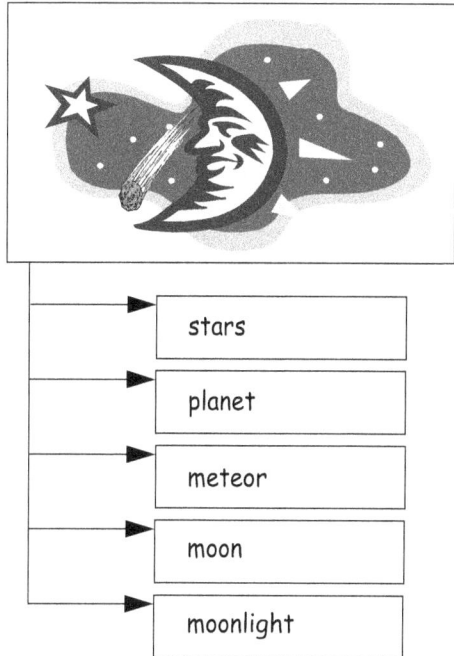

| stars |
| planet |
| meteor |
| moon |
| moonlight |

<u>Do Not Underestimate</u> the value of this as a first step in discovery at every level of knowledge. When you encounter something you have never seen before, start by listing the elements of system.

The accessibility of this "simple" activity to early learners should not diminish its importance at very sophisticated levels in science. This analysis is widely used from medicine to particle physics. For example, it can be used to diagnose the presence of a genetic disease by analyzing the pattern of blood proteins by electrophoresis in a medical laboratory (right).

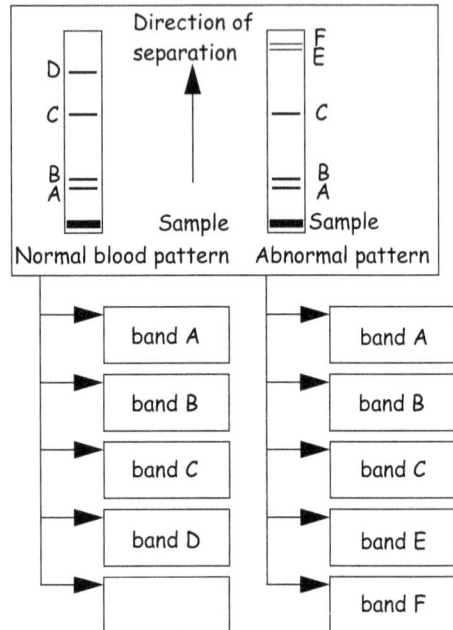

Normal blood pattern    Abnormal pattern

## Element Counter

The basic Element Analyzer can be modified to allow the separated elements to be described in more detail.

The first modification is to count the number of elements in the group. This is the **Element Counter**:

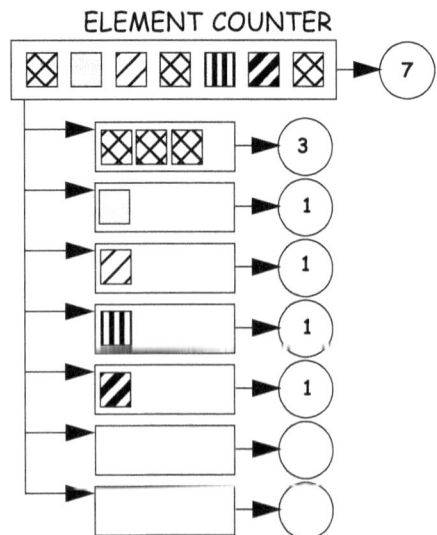

ELEMENT COUNTER

Here is another example of counting the elements of a system with the **Element Counter**:

ELEMENT COUNTER

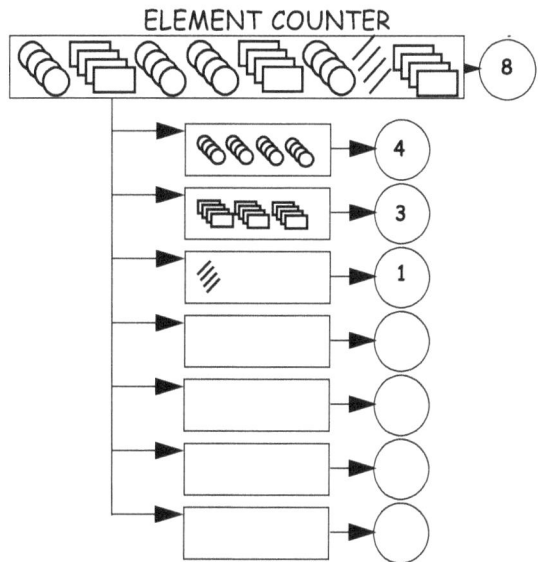

## The Element Extractor

We have already learned that each element in a system has its own structure, that is, it is a sub-system. The Element Analyzer can be extended to further define the subsystems or substructures in a system. This is the **Element Extractor**.

In this example, a single element is being further analyzed. The Element Extractor has extracted three similar squares (**B**) from the overall group of seven (**A**) [the observable - fill pattern - is listed]. The properties of these similar squares are further extracted in the third tier of the analyzer (**C**).

ELEMENT EXTRACTOR

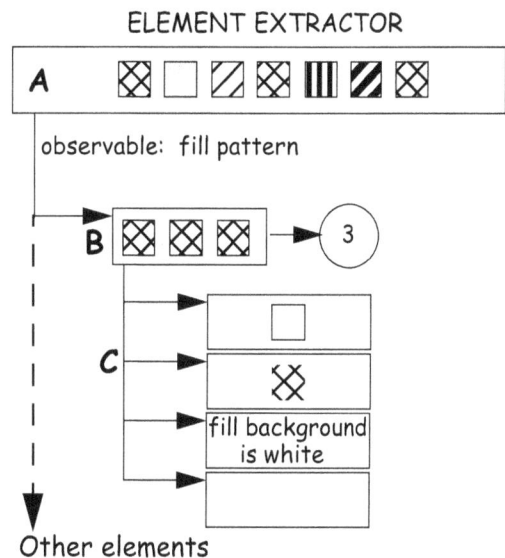

observable: fill pattern

Other elements

In the fourth tier (**D**), the cross-hatch pattern is a system itself and can be further extracted.

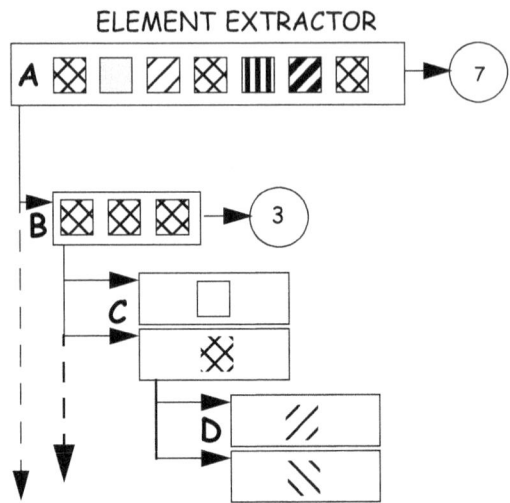

ELEMENT EXTRACTOR

## The Element Analyzer: Single Attributes Practice Exercise

The elements in the systems that we have been looking at have had only two attributes: a uniform geometry and a varying fill pattern. Draw an Element Analyzer when the elements in a system share every attribute (i.e., each element is identical).

**Answer** Your figure could resemble this one:

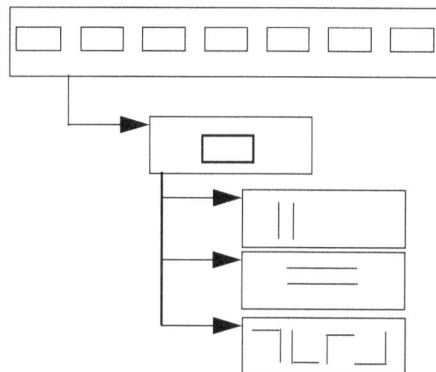

**Single attribute analysis** Below are two examples of a single attribute analysis in different situations. The first is a chemistry example, and the second is speech and drama. Each structure is system made from a monotonous series of elements. In the chemistry examples, the chemical structure in **A** is related to swamp gas, and the chemical structure in **B** is related to animal fat. In the speech examples, the form in structure **A** is an interjection or comment, while the speech structure in **B** is a primal scream.

|  | Structure A | Structure B |
|---|---|---|
| **Chemistry** | (-◇-) | (-◯-◯-◯-◯-◯-◯-◯-◯-◯-◯-◯-◯-◯-◯-◯-◯-◯-◯-) |
| **Speech** | "EE" | "EEEEEEEEEEEEEEEEEEEEEEEEEEEEEEEEEEEEEE" |

This analysis shows how the arrangement of the elements, even when they are the same, can lead to very different properties in systems and structures. This property of repeating elements is a key to understanding organic chemistry, biochemistry, and the repeating structure of crystals in nature.

## Technical Writing Analyzer

All of the properties we have extracted so far can be labelled. If we choose to label them in a written language, the grid would look like this:

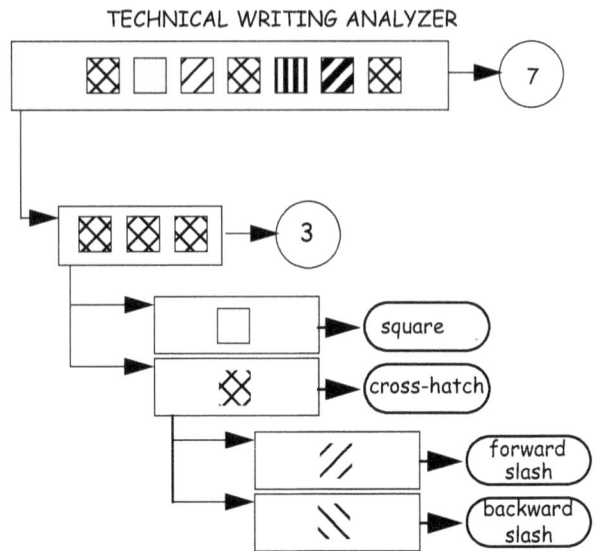

TECHNICAL WRITING ANALYZER

This extension of the Element Analyzer is called a **Technical Writing Analyzer**. The extraction of the information in this highly complex structure should not be underestimated. If this grid is now read in English, a technical summary of the extracted element can be easily and precisely composed:

> In this structure (system) there are seven squares.
>
> Three are filled by a cross-hatched pattern.
>
> The cross-hatched pattern is made from forward and
> backward slashes.

The properties used in the Technical Writing Analyzer do not have to be expressed in English.

## Technical Writing Assistant

Spoken and written language is the first and most common symbolic language used by most people. However, the following example will demonstrate that this process of Technical Writing can be applied for precise analysis and description in any field of interest.

If the expression is a chemical structure, then chemical symbols will translate the structure into the formal language of chemical notation. Two such examples follow.

The first example uses the Technical Writing Analyzer to decode the cleaning fluid, carbon tetrachloride; it might be used in a middle school science class teaching chemistry terminology.

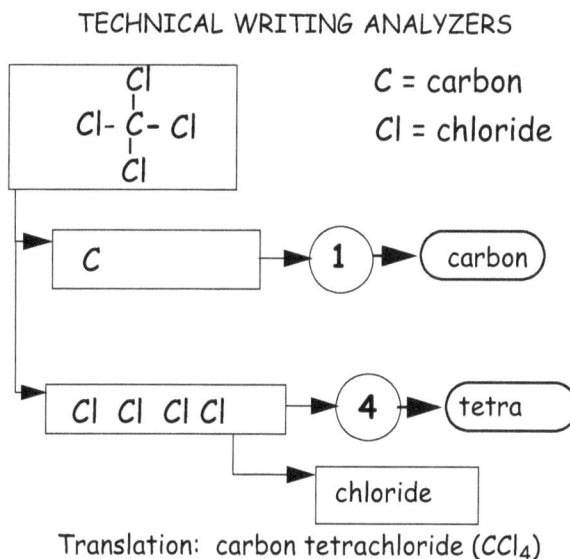

TECHNICAL WRITING ANALYZERS

$$Cl-\underset{\underset{Cl}{|}}{\overset{\overset{Cl}{|}}{C}}-Cl$$

C = carbon

Cl = chloride

C → 1 → ( carbon )

Cl  Cl  Cl  Cl → 4 → ( tetra )

chloride

Translation: carbon tetrachloride ($CCl_4$)

The second example might be seen in a more advanced, University level organic chemistry class. Both use the same process that has already been described

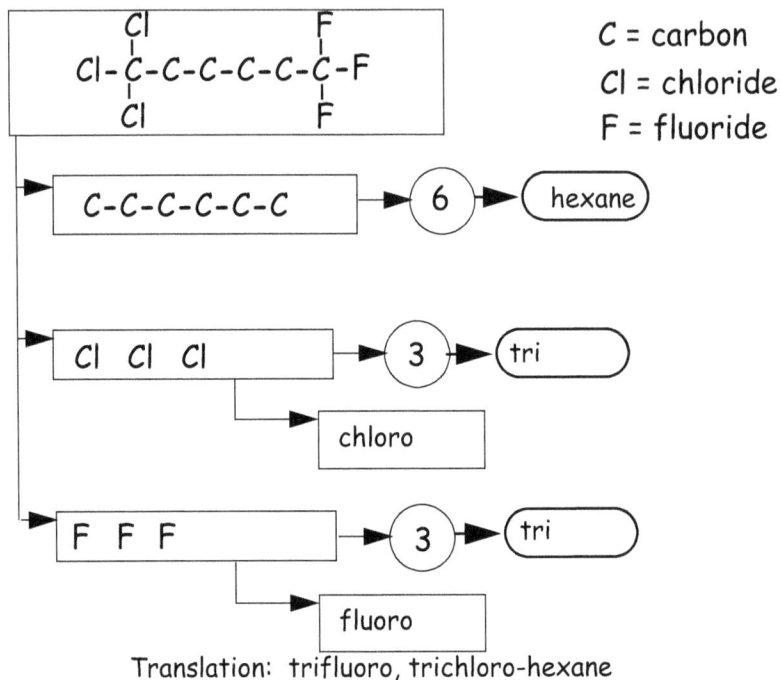

C = carbon
Cl = chloride
F = fluoride

Translation: trifluoro, trichloro-hexane

This is a very useful tool. This process takes an observation through a cycle of critical analysis to provide an expression of the observation in a formal language. This is a process of **formal modeling**. This type of formal modeling is useful for communication of ideas and descriptive data, a vital initial step in the scientific method of inquiry.

**Cross-Curricular Example**

To show the breadth and power of the Technical Writing Graphical Analyzer, a football play is diagrammed. This is a method of annotating the complex movements of 11 players. When a play is called in the football huddle, a cryptic language is used that provides all of the information needed to put every player in place and then into action. In this case, the play is "T-Right-34-Trap". This notation arranges the linemen and backfield players, assigns blocking patterns, and details ball handling instructions to create an opening in the defensive positions. The fullback (the "3" back) is given the ball and runs into this opening.

The players execute the play physically after receiving the verbal description.

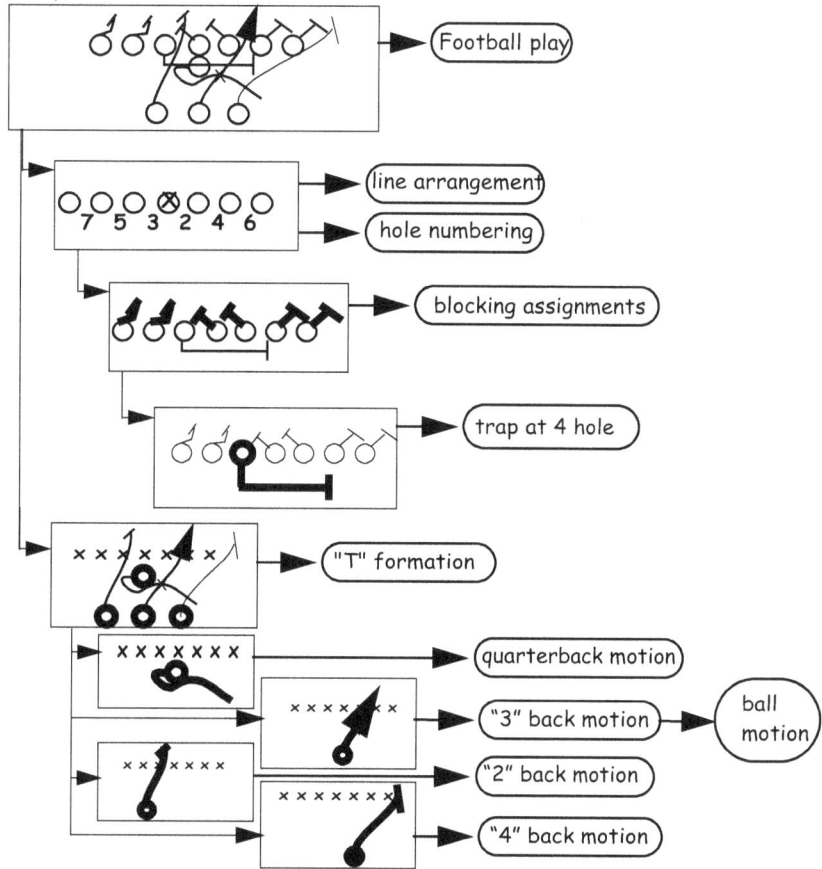

TRANSLATION: T-Right-34 Trap

## Coding Grid

A Coding Grid guides the re-assignment of an element in one system into a symbolic element in a second system. Coding Grids are more directed than the closely related technical writing analyzers that extract and summarize information.

A simple one-to-one attribute exchange can be performed by students as early as first or second grade. More sophisticated translation and coding can be expected as age and education progresses. The Coding Grid guides an essential modeling process of mapping an observation into a symbolic system using the following steps:

CODING GRID

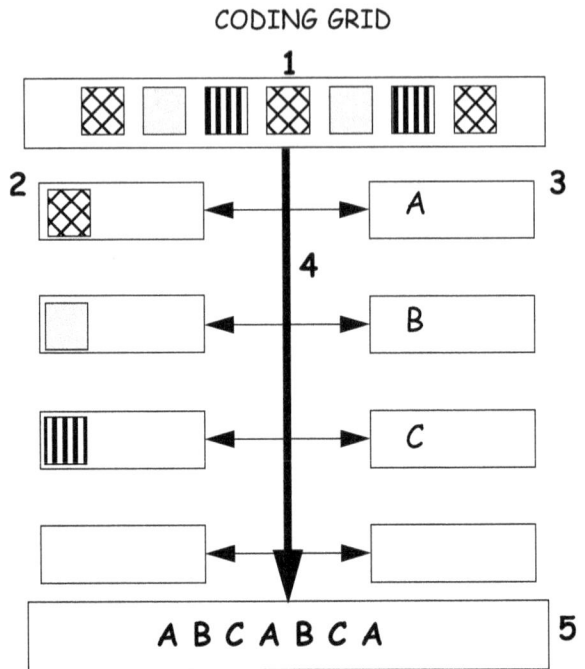

1. The appropriate observables are chosen from the **system** to be modeled.

2. The system is described using the elements, arrangement, rules, and background space.

3. A corresponding set of **symbols** is selected.

4. The observable properties are mapped onto the symbolic forms.

5. The symbols can then be used to represent or **model** the original system.

**Practice Exercise**

Students experimenting with batteries, switches, and bulbs to make circuits can use a Coding Grid to report their work using standard electronic symbols. Complete this Coding Grid to draw a correct schematic for the circuit shown.

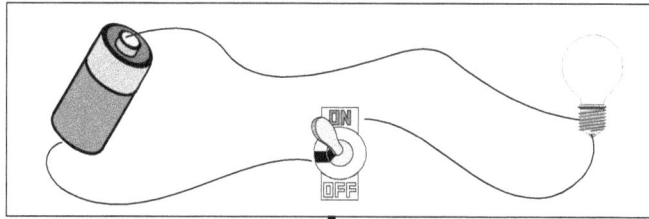

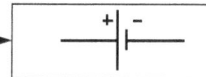

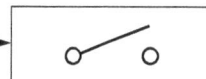

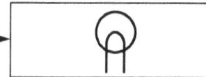

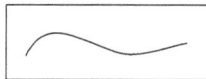

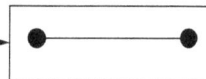

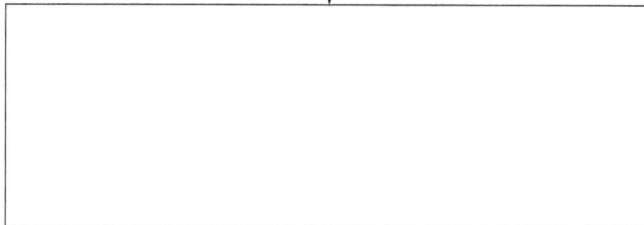

### Answer · The answer is given below:

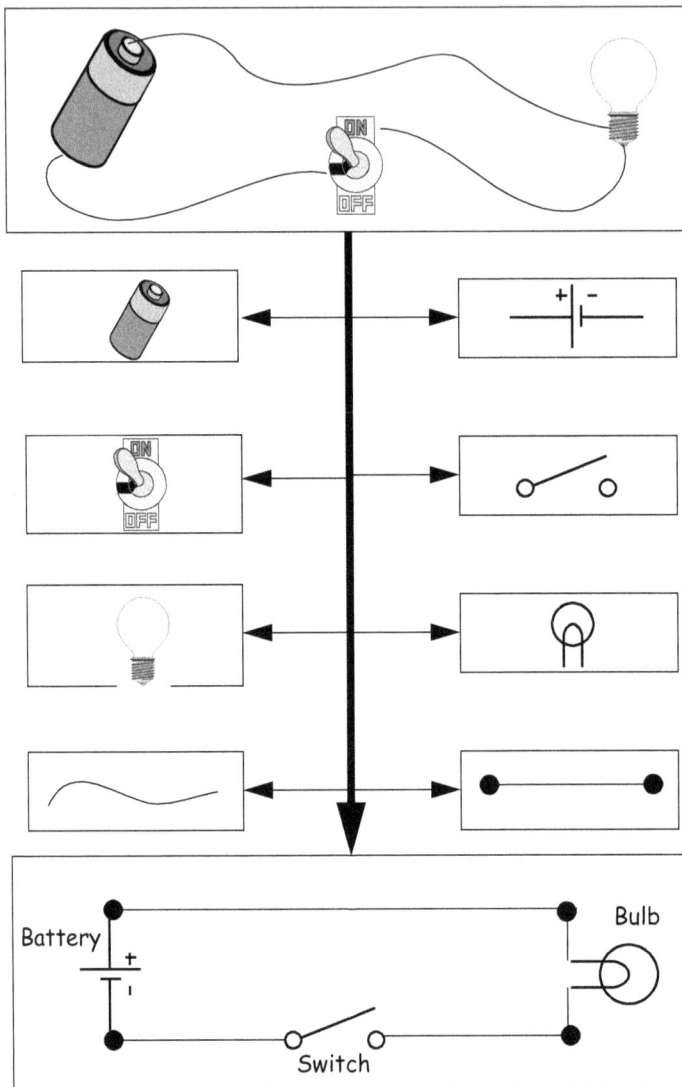

# CHAPTER 17

# Sorting and Classifying with the Graphical Analyzer System

## Observables and Elements

| Analyzers | Application |
|---|---|
| Sorting Analyzer | Sort, group, classify, and sub-sort |
| Multiple Observable Sorting Analyzer | |
| Classifying Analyzer | Make classifying keys and perform logical sorting |

## Sorting and Grouping

When any system or structure of interest is observed, the natural first step is to group and sort the various elements. The grouping and sorting is always based on some attribute of the elements that is chosen by the observer. Often this attribute may not be clearly stated. Therefore, in critical thinking, the first step is to recognize what attribute is being used or observed in order to begin the sorting and grouping process. The chosen attribute is the **observable** used to perform the sort. The Graphic Analyzers presented in this chapter are useful for sorting, classifying, and exploring how observables and attributes affect a classification scheme.

Often grouping and sorting by "obvious" observables does not lead to a complete understanding of a system. In these cases, the scientific approach is to make a more complete exploration of the full set of properties in the system. In the next chapter, we will tour the Graphical Analyzers used for these more extensive investigations.

## Sorting Analyzer: Single-Sort

A system of objects can always be partly described in terms of its elements. Here, a Sorting Analyzer will be used to categorize elements in a system.

The oval space in the center of the Analyzer states the observable. The observable is the property that has been chosen by the observer to evaluate the elements. Here the observable chosen is "shape". The sort that results from choosing this observable is shown. In this instance, shape is probably not a useful choice to describe this system.

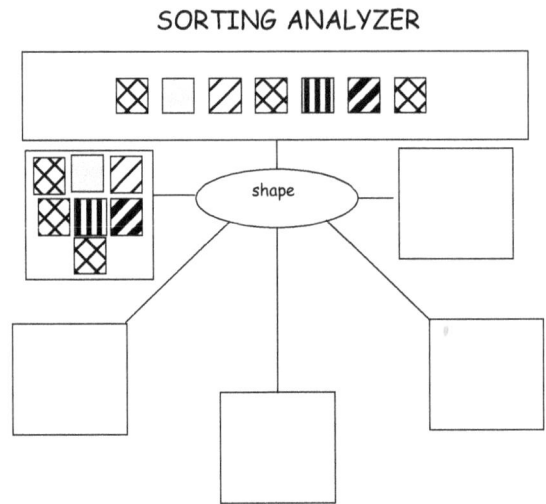

SORTING ANALYZER

Perhaps a better choice of observable would be having "lines inside". The observable of having "lines inside" effectively selects the stippled block from the others. This is a practical result, but its usefulness depends on the questions being asked by the observer.

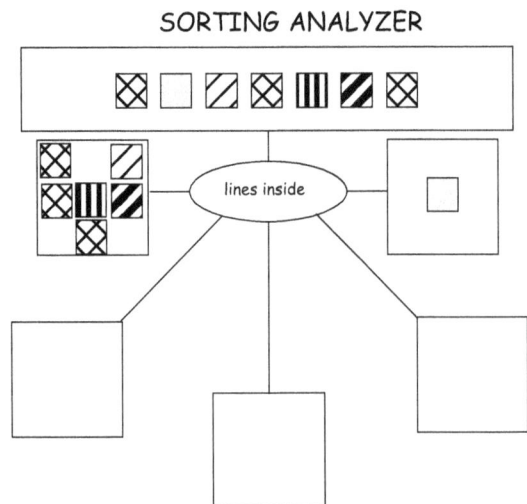

SORTING ANALYZER

If we want to know how many different

kinds of blocks there are, none of the previous observables is useful. Instead we would choose "fill pattern" as the observable.

Using fill-pattern as the observable, the Sorting Analyzer sorts the square pattern system similar to that of the Element Analyzer.

SORTING ANALYZER

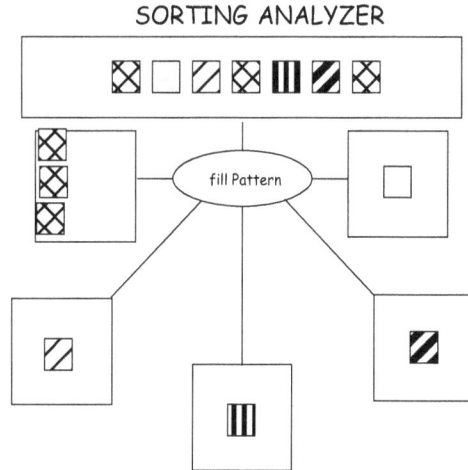

ELEMENT ANALYZER

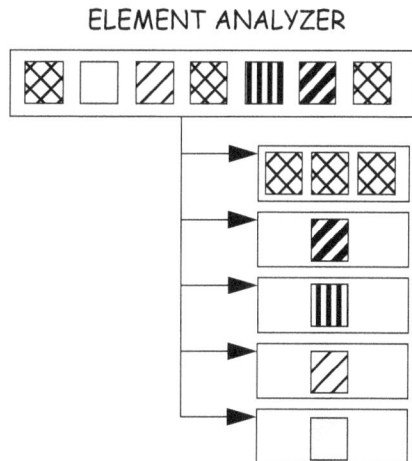

---

**Practice Exercise**   Sort these animals using the labeled observables. (The animals can be identified by name or given a letter symbol to sort.) Animals with the same characteristic of the observable should be grouped together. For example, in the first sort (number of legs), the sorting bins would contain animals with no legs, 1 leg, 2 legs, 4 legs, etc. Possible answers on next page.

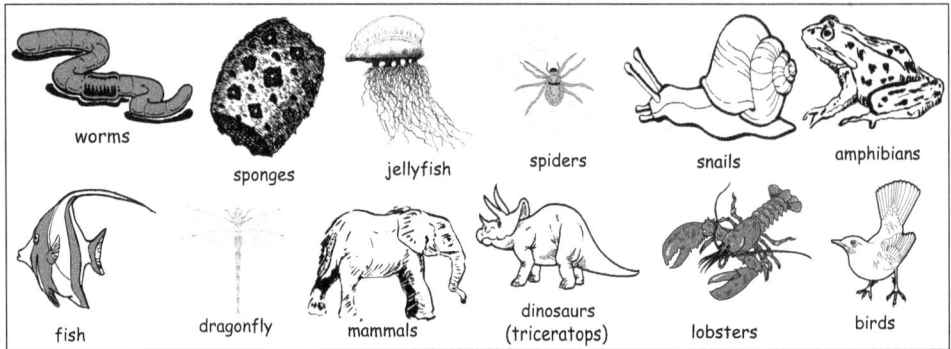

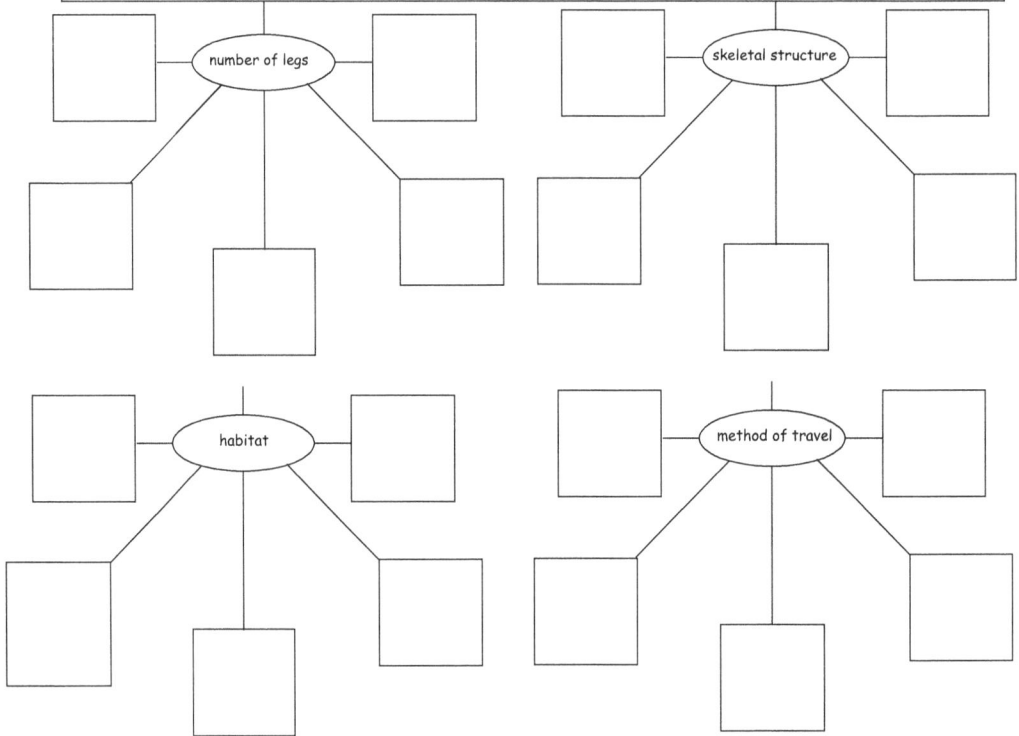

worms

sponges

jellyfish

spiders

snails

amphibians

fish

dragonfly

mammals

dinosaurs (triceratops)

lobsters

birds

number of legs

skeletal structure

habitat

method of travel

**Answer Sorting Analyzer: Poly-Sort**

This is an answer to the animal sorts. If different observables sort the same group at the same time, a poly(morphic) sorter is made. This answer has been drawn as a polymorphic sort.

POLYMORPHIC SORTING ANALYZER

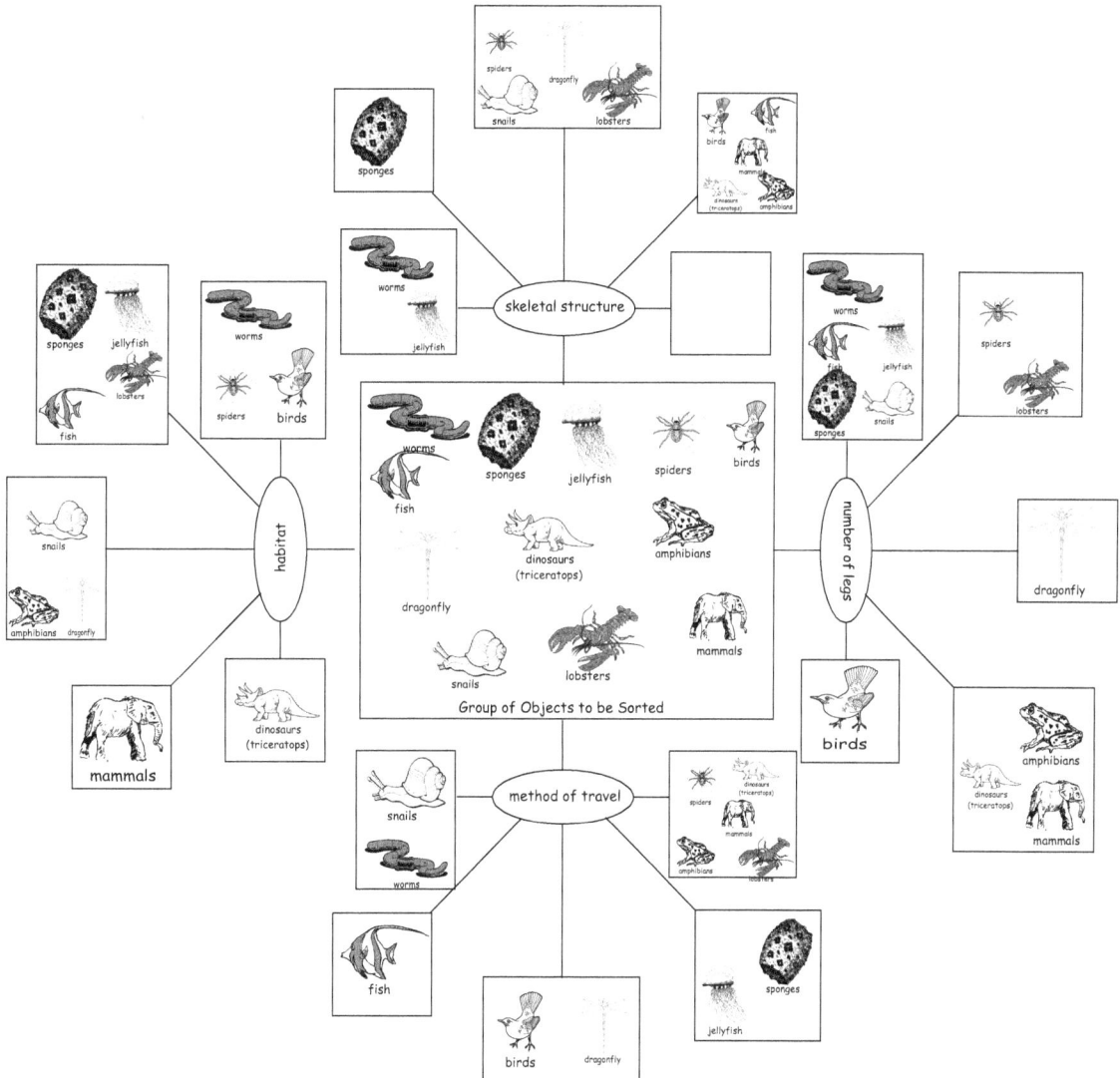

**The Sorting Analyzer: Multiple Attributes**

The purpose of the Sorting Analyzer is to allow a methodical categorization of the elements of a system. Many systems have multiple attributes, which can lead to a variety of sorting strategies. One strategy may be more useful for answering certain questions than another strategy. It is often useful to be able to sort the same group of objects by different observables. This is an important way to search for connections within a group. When searching for connections, each sort pattern can be compared in a Compare and Contrast Analyzer. Consider this system:

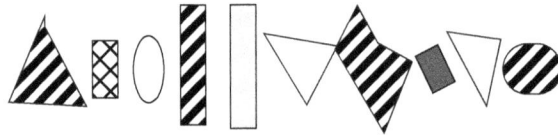

**Practice Exercise**

Try sorting this grouping by the different attributes or observables listed in the observable oval. Possible answers appear on the next page.

SORTING ANALYZERS

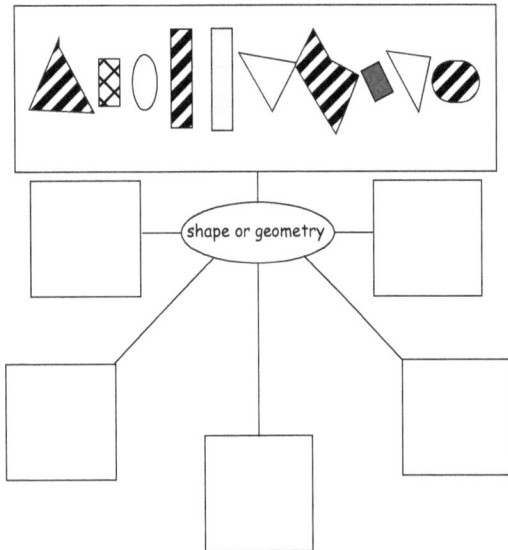

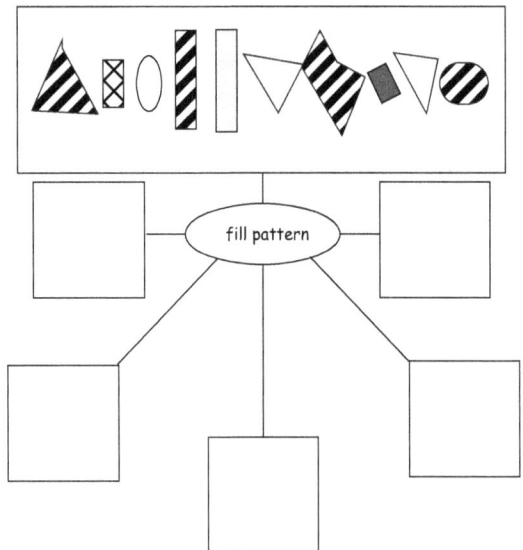

**Answer**   Here are some possible solutions for the previous sorts.

SORTING ANALYZERS

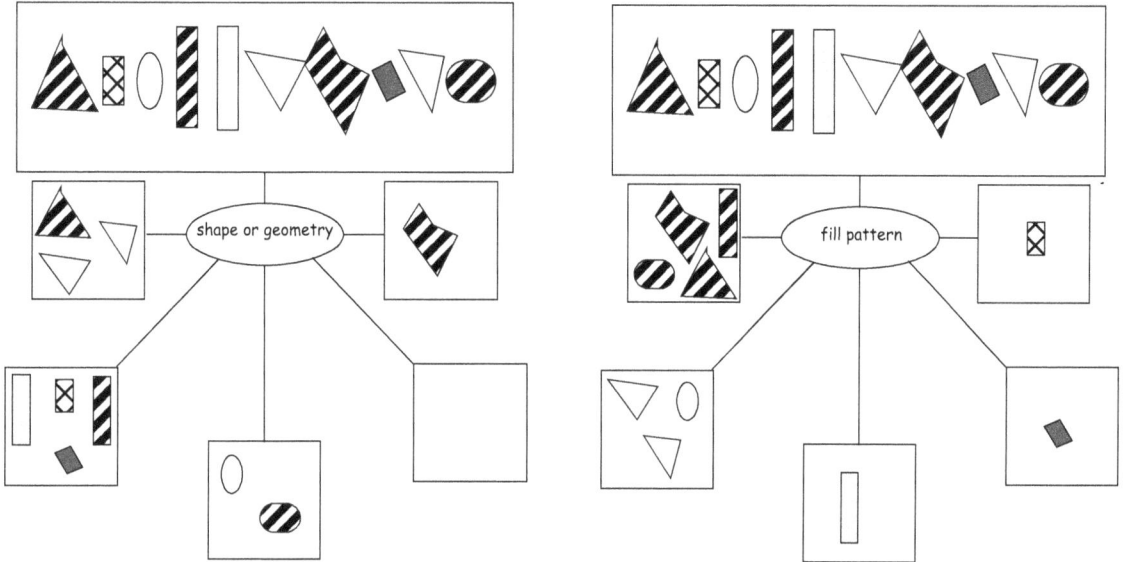

**Practice Exercise**   By reversing the exercise, inference skills can be developed by asking, "What observable would have created this sort pattern?"  This is not as easy as sorting by observable!

SORTING ANALYZERS

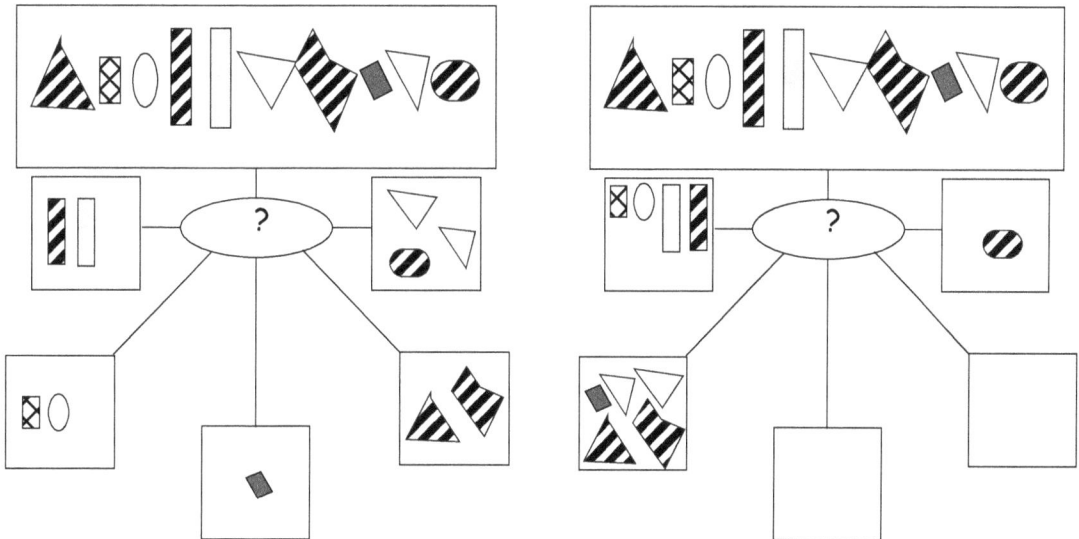

In the exercise on previous page, the observable in sorting analyzer on the left is <u>size</u>; the observable in Sorting Analyzer on right is <u>long axis orientation.</u>

## Sub-Sorting Analyzer

Finally, sub-sorts can be performed on previously sorted elements.

**Practice Exercise**  Try this example.  One of the elements in the sub-sorting bin has been entered.  [Answer on the following page.]

SUB-SORTING ANALYZER

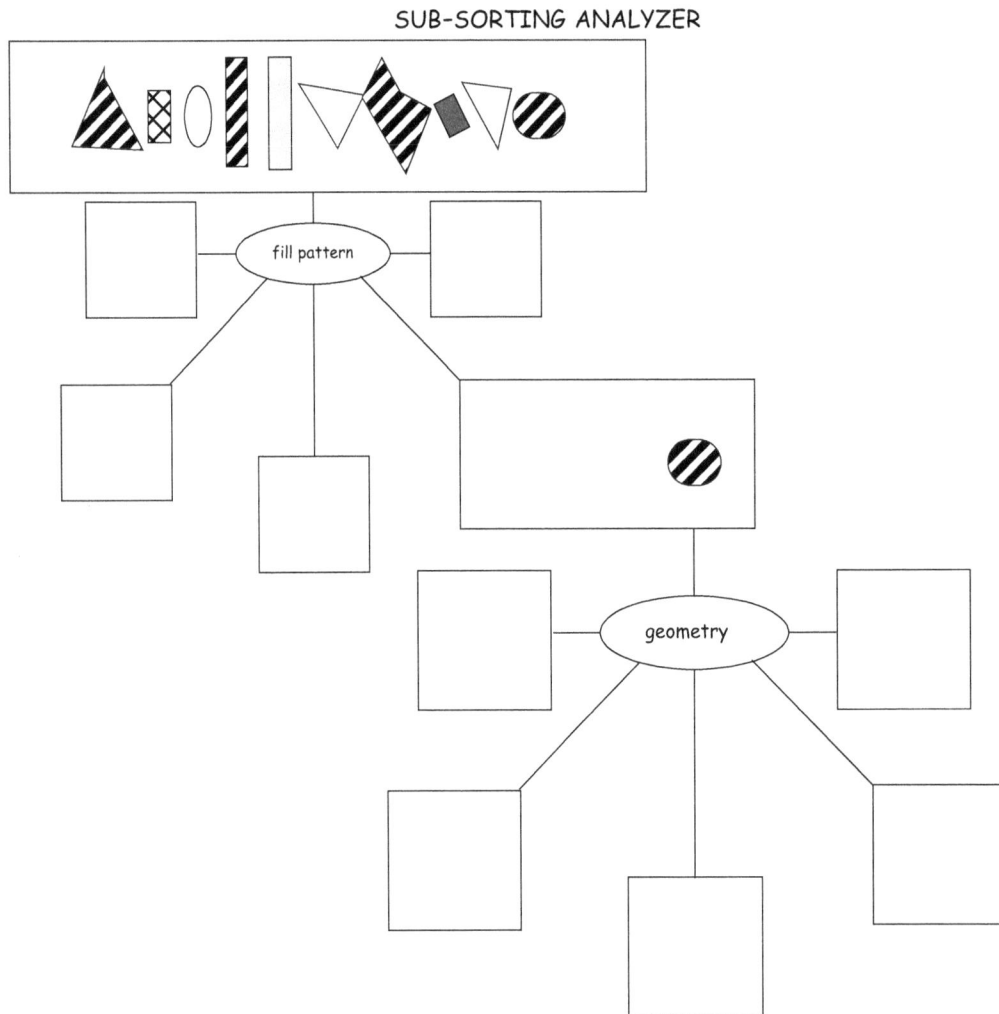

**Answer**   This is a sequential, single observable sort and single observable sub-sort of a set with multiple attributes.

SUB-SORTING ANALYZER

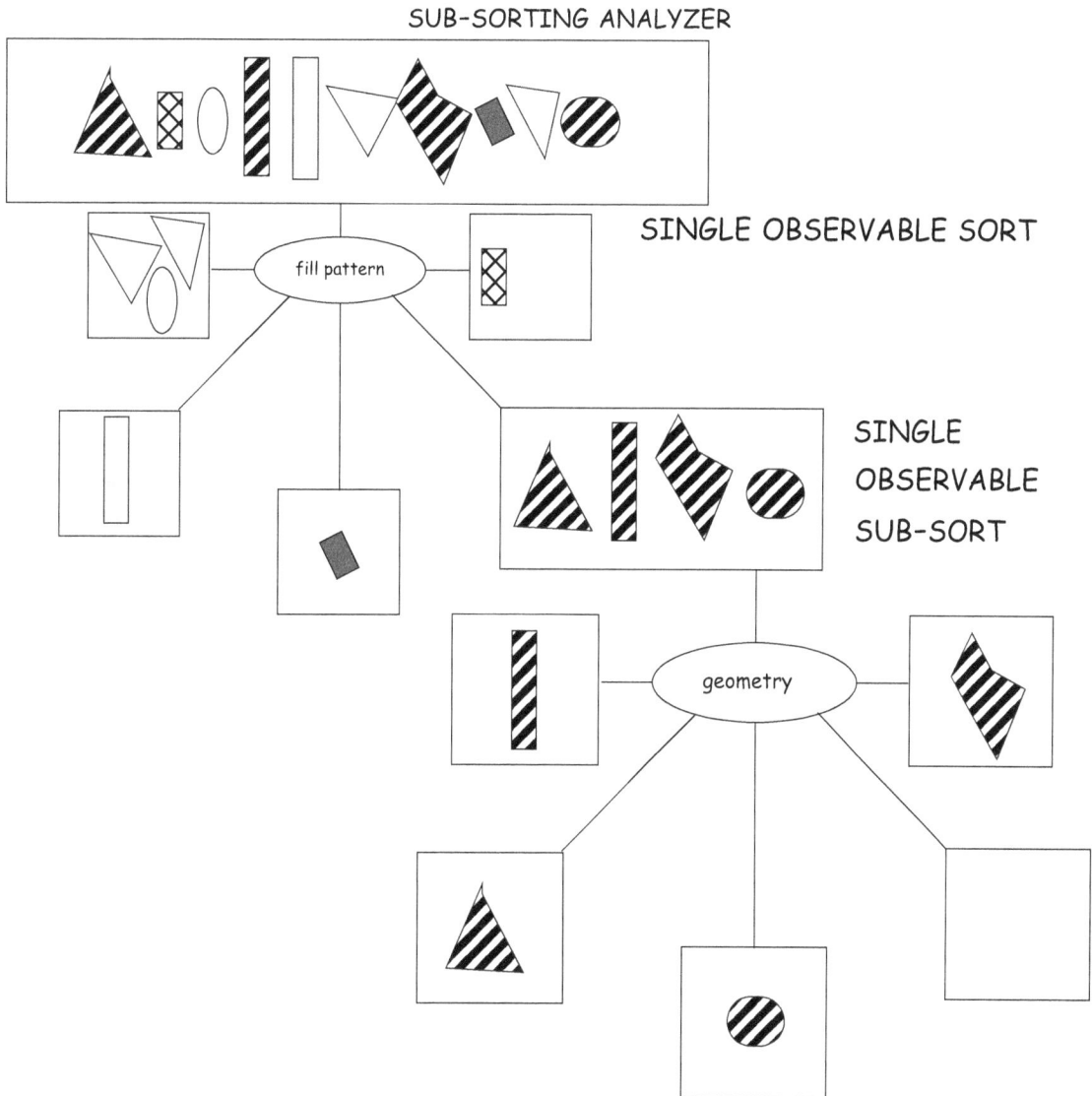

SINGLE OBSERVABLE SORT

fill pattern

SINGLE
OBSERVABLE
SUB-SORT

geometry

**Practice**   Sub-sorting is a very important classification process in scientific
**Exercise**   inquiry.   In this example, chemical elements may be classified by
physical state and by the types of chemical reactions that they
undergo. First, sort this group of chemical materials by their state of
matter at room temperature (25° C). Next sort the gases according to
their pattern of response to flame or fire. For many of us, this
information will not be at our fingertips and may require further
research into the science content of specific chemical reactions.  For
the student, this turns the question of process into a minds-on research
project.

SUB-SORTING ANALYZER

# Chemical Identification

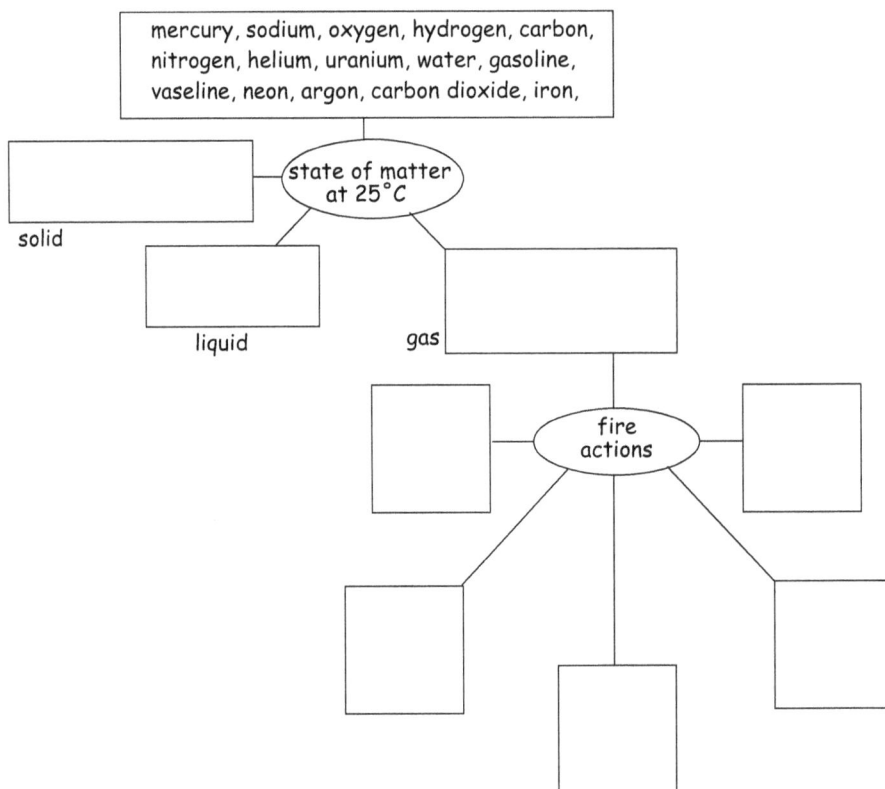

mercury, sodium, oxygen, hydrogen, carbon,
nitrogen, helium, uranium, water, gasoline,
vaseline, neon, argon, carbon dioxide, iron,

state of matter
at 25°C

solid

liquid          gas

fire
actions

## Cross-Curricular Exercise

As in the previous example, sports and games are a natural area for sub-sorting practice. First, separate the list of recreational activities into the labelled game types, an activity may be used more than once at this level. Next separate the groups by the second attribute listed.

SUB-SORTING ANALYZER

Rook, Monopoly, Crazy Eights, Hearts, Soccer, Football, Hide and Seek, Old Maid, Go-Fish, Slap, Bridge, Scrabble, Kickball, Baseball, Softball, Rubric's Cube, Tag, Candy-land, Chutes and Ladders, Hares and Hounds, Anagrams, Kick the Can

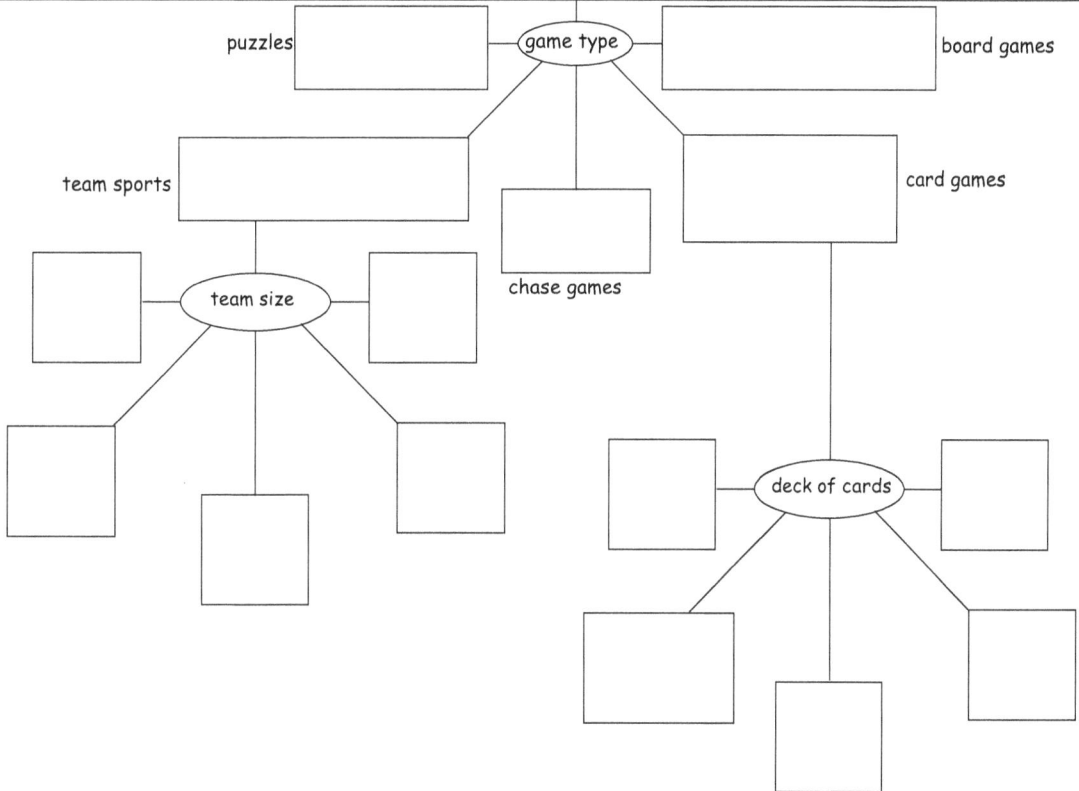

puzzles

game type

board games

team sports

card games

team size

chase games

deck of cards

**Answer**   Answers to the previous sub-sorts.

## Chemical Identification

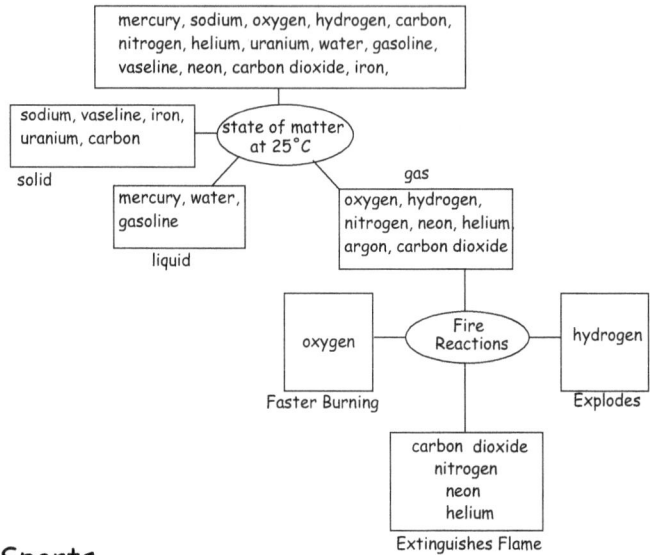

mercury, sodium, oxygen, hydrogen, carbon, nitrogen, helium, uranium, water, gasoline, vaseline, neon, carbon dioxide, iron,

sodium, vaseline, iron, uranium, carbon

(state of matter at 25°C)

solid

mercury, water, gasoline

liquid

gas

oxygen, hydrogen, nitrogen, neon, helium, argon, carbon dioxide

oxygen

(Fire Reactions)

hydrogen

Faster Burning

Explodes

carbon dioxide
nitrogen
neon
helium

Extinguishes Flame

## Games and Sports

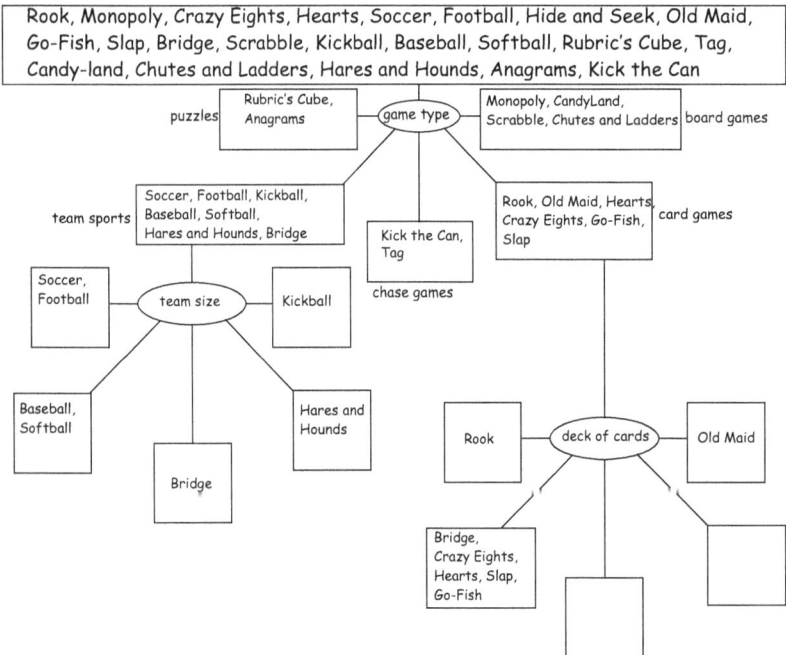

Rook, Monopoly, Crazy Eights, Hearts, Soccer, Football, Hide and Seek, Old Maid, Go-Fish, Slap, Bridge, Scrabble, Kickball, Baseball, Softball, Rubric's Cube, Tag, Candy-land, Chutes and Ladders, Hares and Hounds, Anagrams, Kick the Can

puzzles

Rubric's Cube, Anagrams

(game type)

Monopoly, CandyLand, Scrabble, Chutes and Ladders

board games

team sports

Soccer, Football, Kickball, Baseball, Softball, Hares and Hounds, Bridge

Kick the Can, Tag

chase games

Rook, Old Maid, Hearts, Crazy Eights, Go-Fish, Slap

card games

Soccer, Football

(team size)

Kickball

Baseball, Softball

Bridge

Hares and Hounds

Rook

(deck of cards)

Old Maid

Bridge, Crazy Eights, Hearts, Slap, Go-Fish

## Classifying Analyzers

The Sorting Analyzer can be modified to form a **Classifying Analyzer** [sometimes these are called dichotomous or binary keys].

The mental process of sorting and grouping tests objects for "sameness". If two things are the same according to the observable or property used in the comparison, then they can be grouped together.

The use of the Sorting Analyzer to sort from a general group has been demonstrated. The next step is to see how the Analyzer can be used to separate and classify objects out of a group. This analytical process is very important throughout scientific classification. It also has an important role in problem solving, logic, mathematics, law, and all forms of art criticism.

Here is how a Classifying Analyzer can be made:

1.  First, the Sorting Analyzer is designed to separate the larger set into two groups.

SORTING ANALYZER

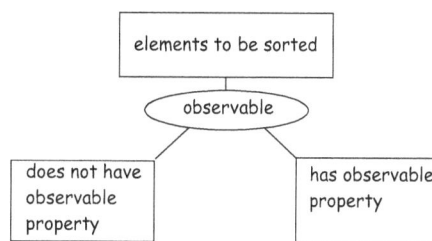

2.    Then, a series of Sorting Analyzers continue to sequentially sort each tier into smaller subgroups. The observable is chosen to precisely separate the elements of each set in a way that will lead to useful subgroups. The ultimate goal is a specific and unique set of observables that can be applied to identify any single member of the set.

CLASSIFYING ANALYZER

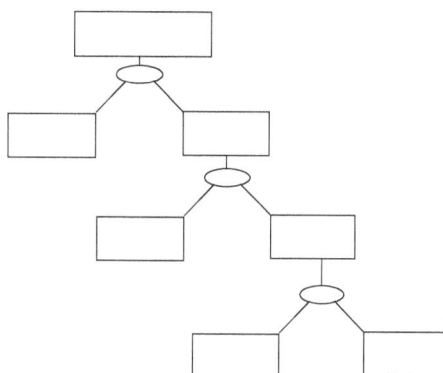

3.    The objects that have the observable property are always placed in the box to the right.

**Practice Exercise** Complete this Classifying Analyzer by sorting and categorizing the elements from the top structure. **The objects that have the observable property are always placed in the box to the right.**

CLASSIFYING ANALYZER

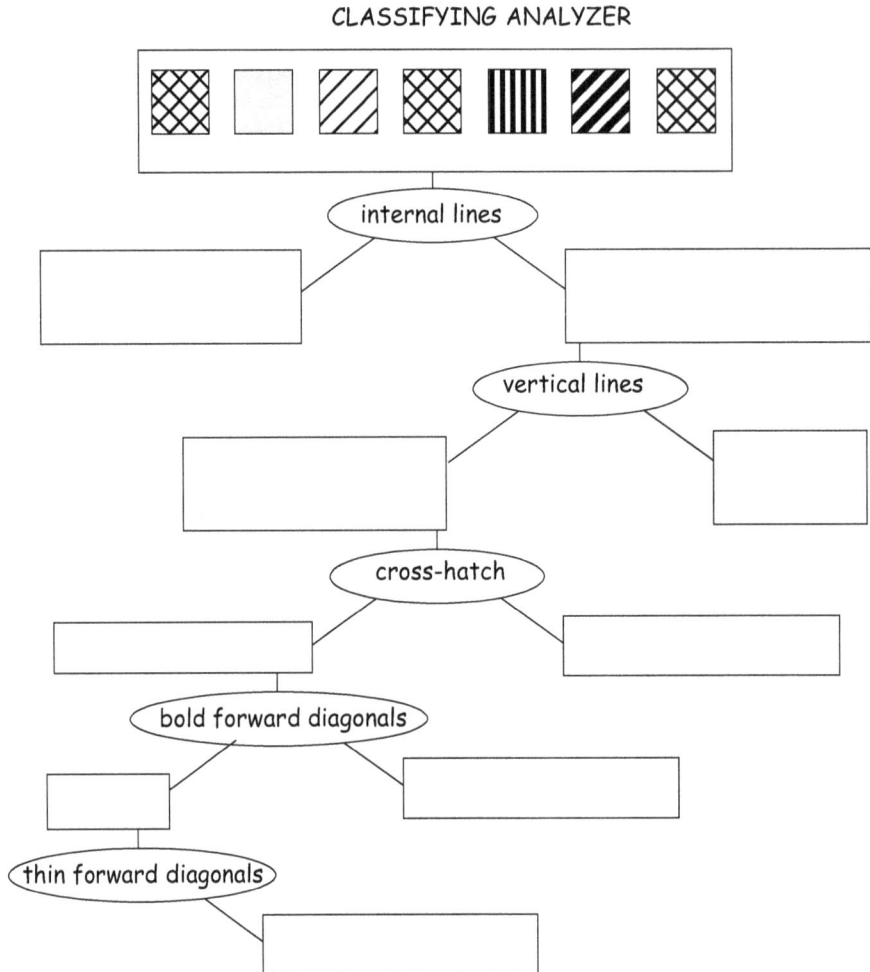

A Teacher Prepares: From Brain To Science Literacy

**Answer**  Here is the answer to the previous exercise. The objects that have the observable property are always placed in the box to the right.

CLASSIFYING ANALYZER

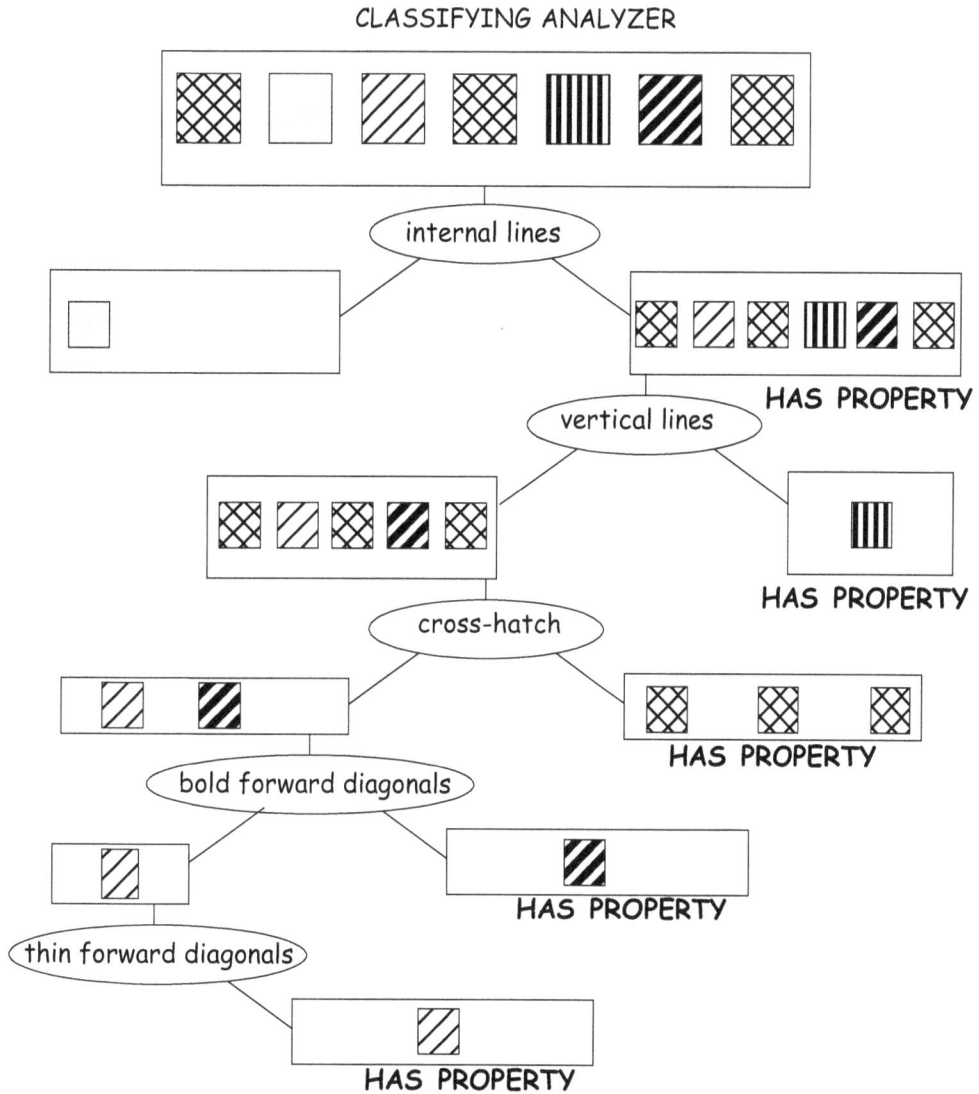

**Practice Exercise** The next example requires some content knowledge of clouds and the weather. Sort the following set of clouds by the observables of precipitation (<u>rain</u>, <u>thunderstorms</u>, <u>steady rain</u>, <u>drizzle</u>). Place the correct observable in the central oval. Place the cloud letter in the sorting bin. The element with the chosen observable is placed in the bin to the right.

## CLASSIFYING ANALYZER

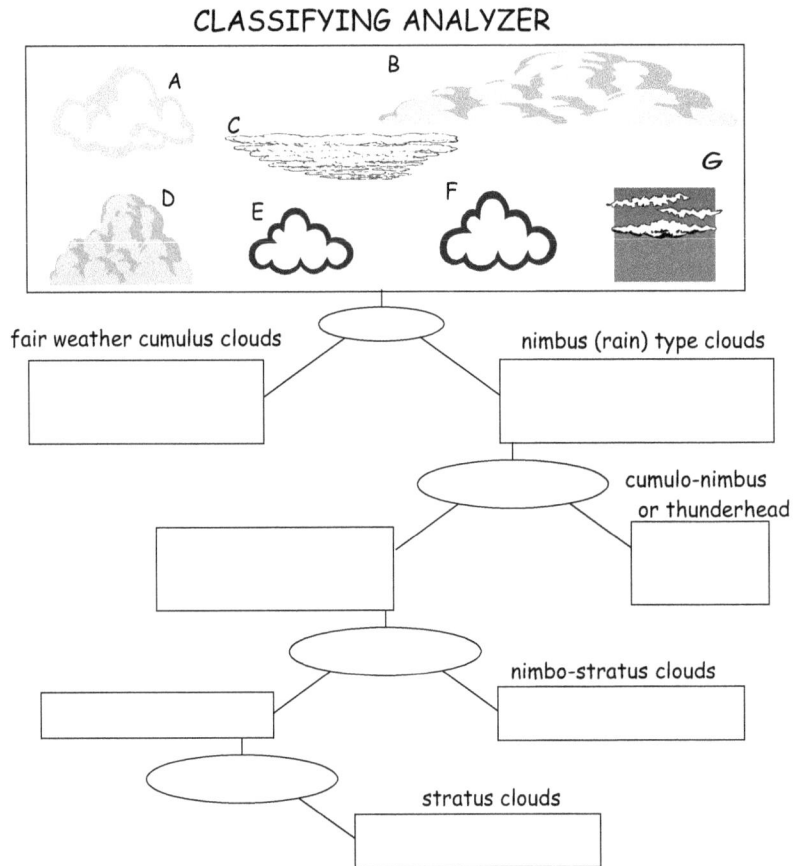

fair weather cumulus clouds

nimbus (rain) type clouds

cumulo-nimbus or thunderhead

nimbo-stratus clouds

stratus clouds

**Answer** This answer to the previous exercise shows how the clouds can be sorted and categorized using precipitation patterns. **Clouds that have the observable property are placed in the right hand box.**

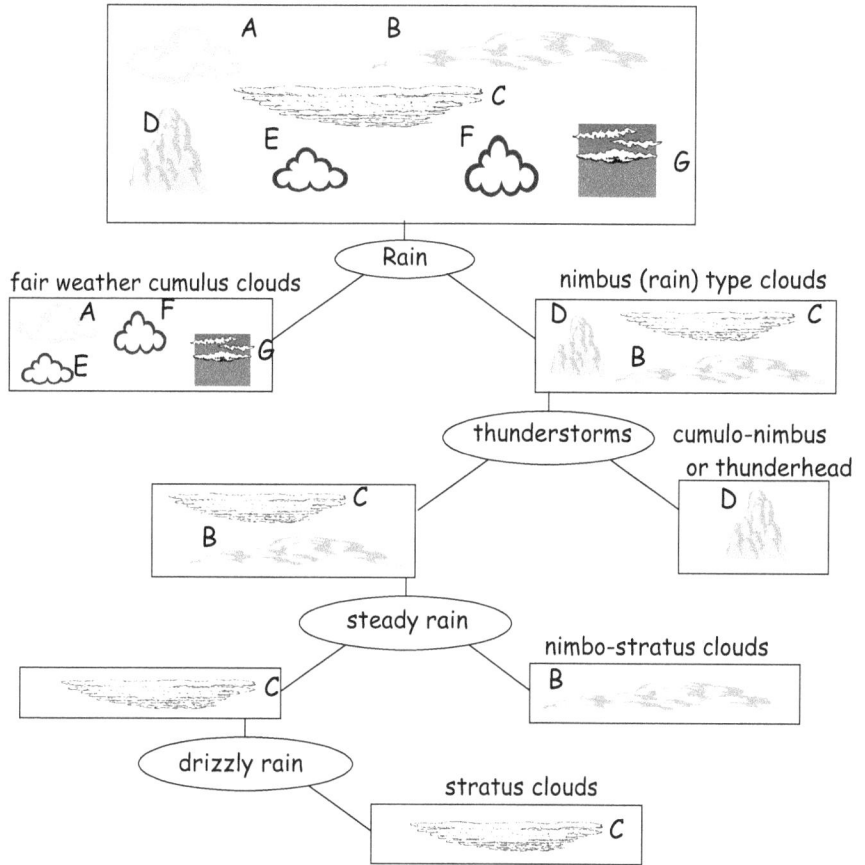

**Cross-Curricular Example**

A Sorting Analyzer can be used outside the science content curriculum. Complete this Category Analyzer using observables inferred in the verse. A variety of answers are possible depending on the initial observable chosen in your analysis.

CLASSIFYING ANALYZER

Jack Sprat could eat no fat
His wife could eat no lean
And so between the two of them
They licked the platter clean.

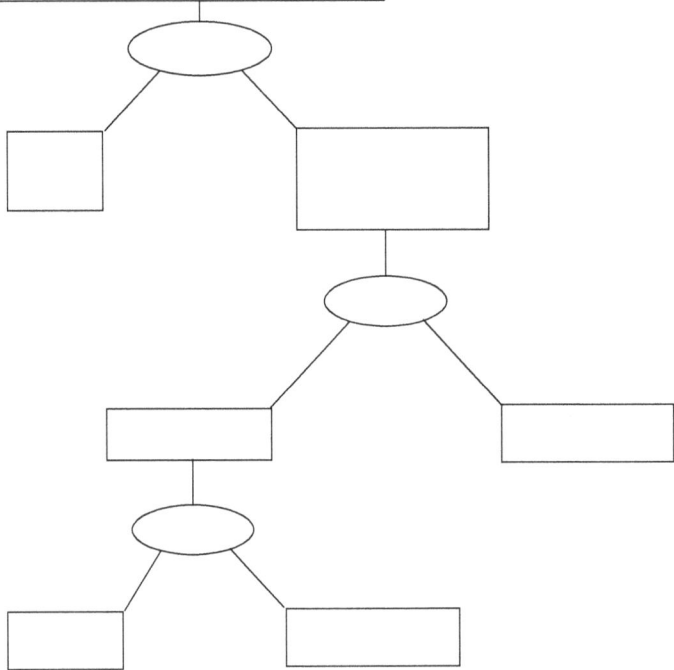

**Answer**   A possible answer to the Jack Sprat verse is given.  Your version may be different depending on the observables that you have chosen.

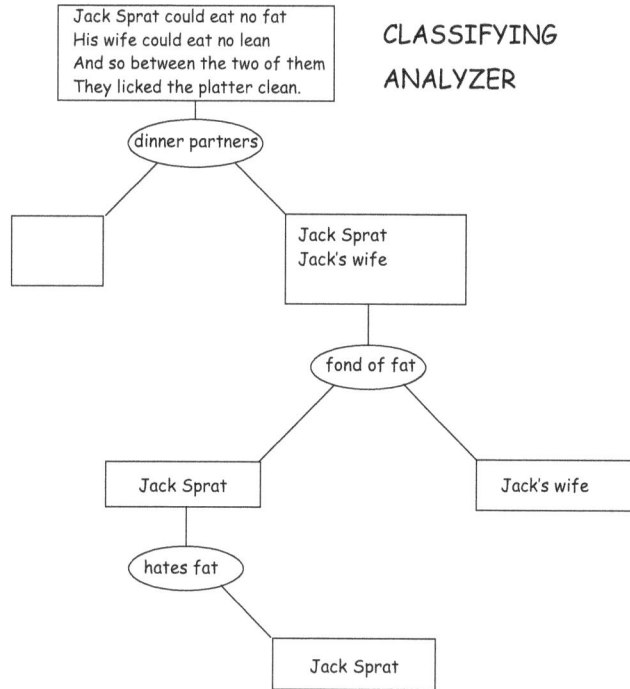

Jack Sprat could eat no fat
His wife could eat no lean
And so between the two of them
They licked the platter clean.

CLASSIFYING

ANALYZER

(dinner partners)

Jack Sprat
Jack's wife

(fond of fat)

Jack Sprat

Jack's wife

(hates fat)

Jack Sprat

**Summary**   In the last two chapters, we have seen how the properties can be used to describe and analyze many aspects of the world.  Now we will turn our attention to describing the background space of a system.

# CHAPTER 18

# Background Space Graphical Analyzer System

| Analyzer | Scientific Process |
|----------|-------------------|
| Background Space Properties Analyzer | Identify the context in which a system is found |
| Background Space Feature Extractor | Define the coordinates and dimensions of a system |
| | Extract the properties of background space in order to infer and describe rules in a system |

## Appreciating the Background Space

The most neglected aspect in exploring the world is an appreciation of the importance of the background space where the elements of a system are found. The background space provides the context or environment of a system. As such, it is a very important aspect of the ecology of a system. Ecology in this context means the overall interaction of each part of a system, not only with other parts of the same system, but with other systems that may also reside in the same background space as well. If the background space is not appreciated, the interaction between systems will not be able to be detected or recorded.

People frequently talk about "ecology" without realizing the importance of knowing the details of the background space. Many people would consider that the concept of ecological thinking was raised to the popular mind by books such as Rachel Carson's *The Silent Spring*. *The Silent Spring* described the damage done to birds and other animals by the use of pesticides, notably DDT, that were used to control the destruction of agricultural crops by insects. In the original frame of reference, the background space for the interaction between insects, plants, and humans focused on the role insects played in the destruction of crops and the transmission of disease. A powerful pesticide like DDT offered benefits to the inhabitants of the Earth. However, it was originally unappreciated that pesticides could damage other species in the biosphere, because it was not recognized that the background space also included the ground water cycle and fish! This is a classic case of discovering the background space due to an unexpected cause-and-effect result that moved through a previously unobserved part of the system.

Biosphere contamination by pesticides is not the only example of unexpectedly confronting the properties of the background space. Many scientific advances result from the new perspective that recognizing the previously unobserved properties of the background space can provide.

## Making Connections Requires Knowledge of the Background Space

To fully describe a structure, we must be able to extract and describe the rules of arrangement. However, rules can only arrange elements "somewhere". Before extracting the rules, we must know the characteristics of that "somewhere".

The common sense of this is obvious, yet every teacher should be aware that the most common mistake made by both the beginner and the experienced scientist is neglecting to perform a careful description of the background space and to pay proper attention to the properties of that space. One of Einstein's great contributions to modern knowledge (general relativity) was simply the recognition of the properties of the background space that makes up our universe.

## Learning About Background Space

Background space is the place where things are put. Learning about the background space has three levels.

   1.   The first level is easy and accessible to every student even in the earliest grades.

- The Progression of Inquiry and Language of Patterns encourages every critical thinker to start thinking about boundaries and the properties of the space where things happen.
- Most early knowledge of background space will come from considering common experiences like boundaries of playing fields, "inside" and "outside" voices, classroom versus playground behaviors, etc.

**2.** The next level is the characterization of the background in more abstract terms, including appreciating that the space is one, two, three, or four dimensional (time matters!). Characterizing time in the background space will be discussed in more detail in the next section: Characterizing Background Space.

- This type of characterization is more sophisticated and requires more math and language skills and a developing capacity for inference and abstraction.

**3.** Finally, a detailed description and mapping of the background space requires sophisticated knowledge and more advanced cognitive skills levels. Students must be able to think metaphorically and use mathematical ideas.

All three levels use the Graphic Analyzer System, but the role that the associated Analyzers will play in the educational setting or your work will depend greatly on the student that you teach or the problems being solved. The crucial recognition of the <u>existence</u> of the background space derives from thinking about it in the first place. This alone is an accessible advance in knowledge for every student at every level!

# Background Space Analyzer

The background space includes three parts:

1. A **sample space** where the system itself is found

2. A **boundary** that defines the extent of the sample space

3. The **surrounding space**, which is the remaining background space outside of the boundary.

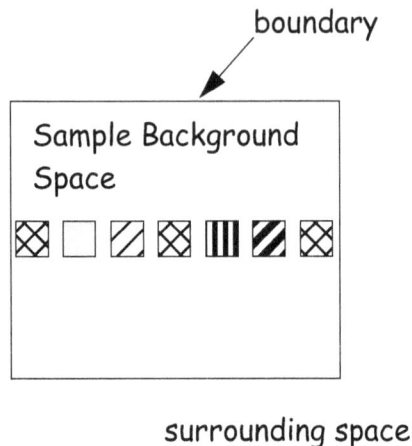

BACKGROUND SPACE PROPERTY ANALYZER

boundary

Sample Background Space

surrounding space

# Describing The Background Space

Background space has properties. These properties can be treated just like systems and elements; they can be analyzed with a Property Analyzer.

Here are a variety of examples to explore the background space and its influence on understanding a system.

**Example 1**

What is the description of the background space for a balloon if it is filled with helium?

- The skin of the balloon is the **boundary** between the outside and the inside of the balloon.

  The properties of the boundary are that it will not let the helium gas across the boundary.

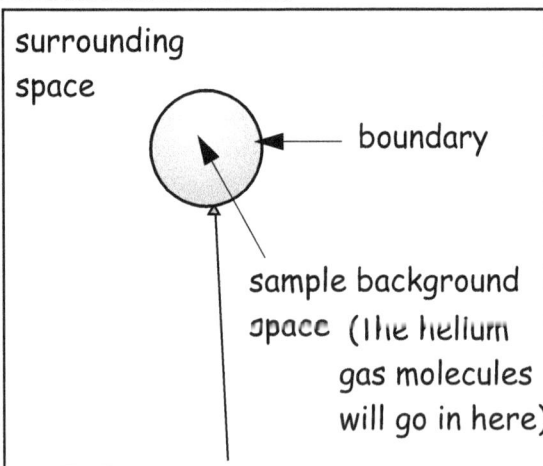

surrounding space

boundary

sample background space (the helium gas molecules will go in here)

- The **sample background space** or container space is the space inside the balloon. The **surrounding space** is the space filled by the air outside the balloon.

The properties of the floating balloon depend on all three of the conditions of the background space. As long as the boundary contains the helium gas within the balloon, the inside sample background space will be filled with less matter than the outside atmospheric space; the balloon will be lighter than air and thus will float. The properties of the space inside the balloon and outside are the same, however, the elements that occupy the inside space are relatively less dense than the outside space.

If the balloon bursts, the boundary is disrupted. The helium gas freely escapes from the **sample space** into the **surrounding space**, and the balloon will no longer be less dense compared to the atmosphere. The skin of the balloon will fall from the sky. This explanation of why a helium balloon rises requires an understanding of the background space in which it exists.

**Example 2** Learning about backgrounds is easier by considering spaces and boundaries that are commonly appreciated.

The sample background space, boundary, and surrounding space of a playing field are labelled. Are the elements treated differently when they are in these different spaces?

surrounding space

boundary

sample background space

The behavior of a player, whether on the field or on the bench, is determined by the boundary of the background space (the sideline).

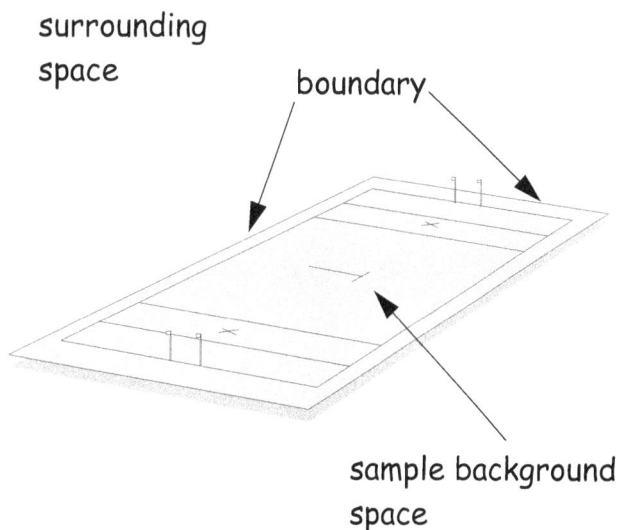

**Practice Exercise**  Label the background space (surrounding space, sample background space, and boundary) for this balloon.

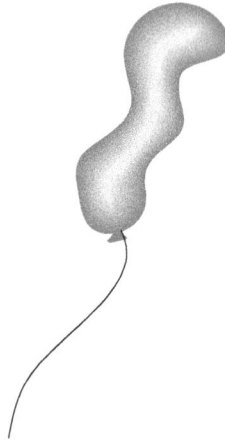

**Example 3**  The parts of the background space for a can of soda in a grocery cart are labelled.

- The overall background space is the space and time in which we all exist. It surrounds the shopping cart.

- The can of soda is arranged in the sample background space of the grocery cart.

- The boundary is the wall of the shopping cart.

- Consider the can of soda inside the sample background space compared to its relationship to the buyer when it sits on the shelf in the surrounding space. Inside the sample space (the shopping cart), the can of soda moves out of the store as a purchase by the shopper, a distinctly different state of existence compared to staying on the shelf in the surrounding space.

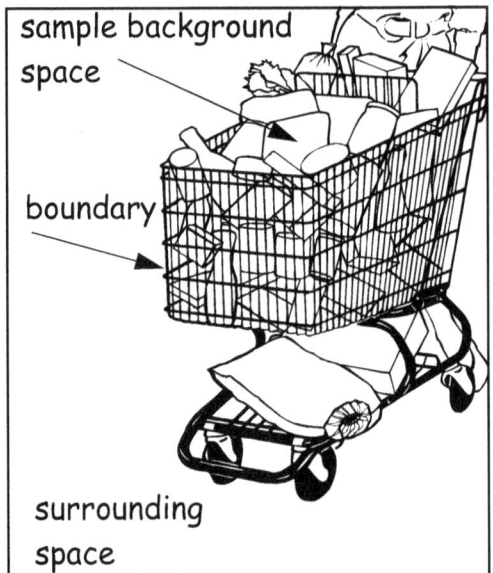

sample background space

boundary

surrounding space

- The properties of each of these parts of the background space play a role in the behavior of the elements in the shopping cart and their relationship to one another.

The previous analysis helps us determine the qualities of the background space and helps us define what the boundaries of interest are in a system. A qualitative description of the space of the system, the surroundings, and the boundaries are all very important in understanding a system's behavior.

**Practice Exercise**  Label the background space (surrounding space, sample background space, and boundary) for this squash court.

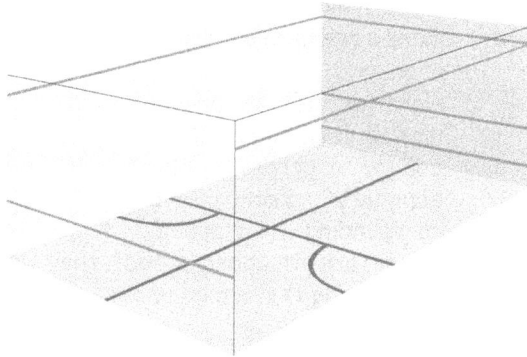

**Example 4** The perpetual motion machine is an example of a background space misconception.

Perpetual motion machines appear to violate the law of nature that says that energy can not be created. Measurements of energy production and use inside the sample space make it seem that more energy is being produced than used. However, all perpetual motion machines draw energy _across_ the boundary from the surrounding space.

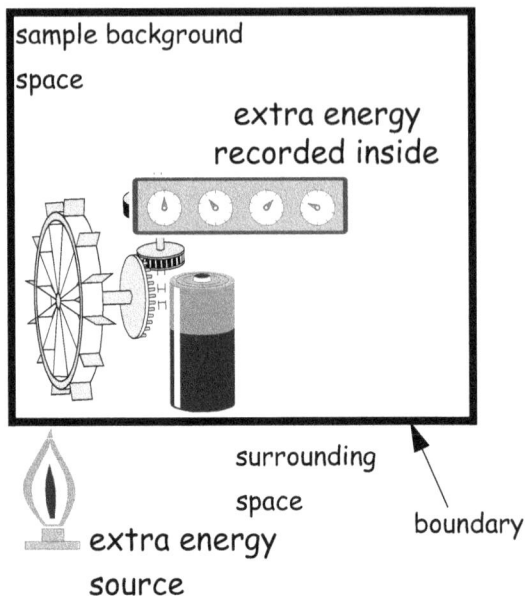

sample background space

extra energy recorded inside

surrounding space

boundary

extra energy source

Therefore, it appears that they are _making_ energy when in fact they are _getting_ it from an unobserved part of the background space.

**Answers to Previous Exercises**

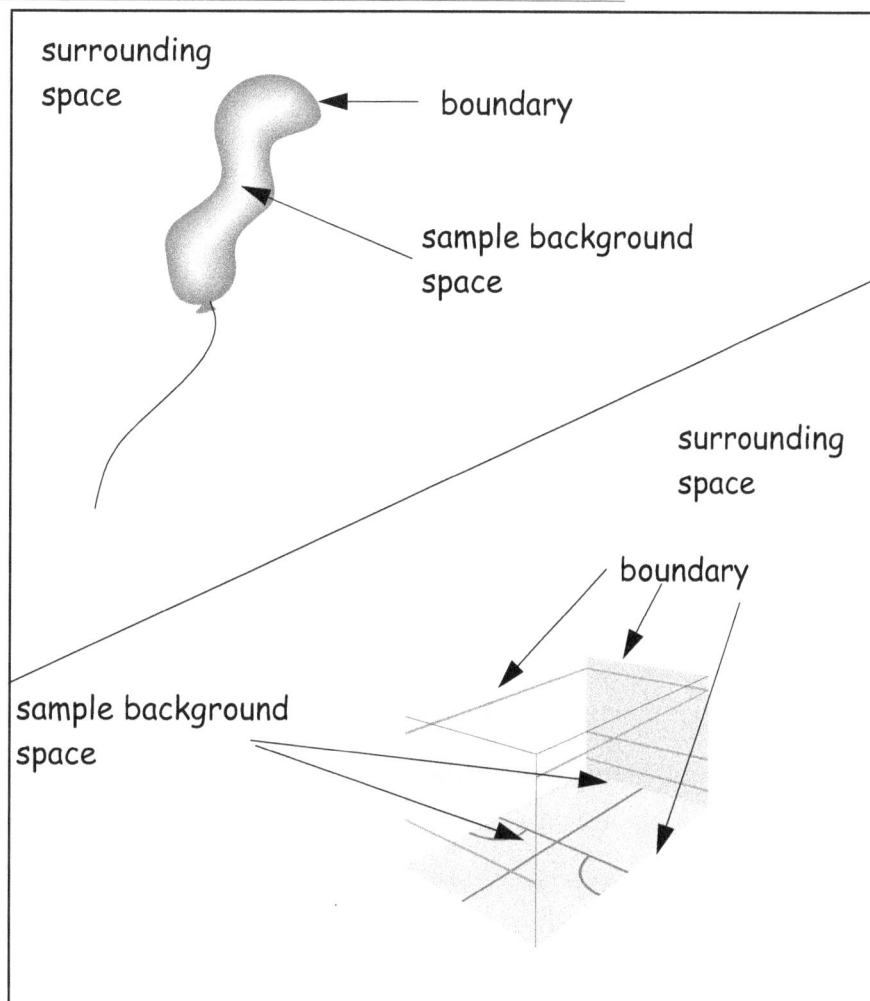

surrounding space

boundary

sample background space

surrounding space

boundary

sample background space

## Sample Background Space

In order to determine and describe the character of the background space with the precision necessary to form a complete map, we must examine the more restricted portion of the background space (the **sample background space**) that contains the structure of interest.

- The easiest way to do this is to define a "container" space. The container holds the system of interest.

· The container's properties are then extended to the rest of the space. In this fashion, we can attribute the properties of the sample background space to the greater background space.

A description of the shape of the space and a map of the space is what a Background Space Analyzer provides.

## Characterizing Background Space

To characterize background space, we must:

1.   Figure out the shape of the space that contains the elements. For example:

· A page in a book (the space in which these words appear) is flat with length and width.

· The Earth's surface (the space on which towns and landmarks are found) is a curved spherical surface.

2.   Have a sense of the size of the sample and surrounding space.

· A system or structure is located in a background space that is large enough to contain it.

If the system is a solid object like an apple, then the background space is at least three-dimensional (3-D). Words or drawings are two dimensional, since they are contained in the plane of the page (having **length** and **width** only).

Almost every system of real interest has a time dimension (a past, present, and future). For a system to undergo change or evolution, it must move through time. Therefore, most background spaces are actually four dimensional (**length, width, height,** and **time**). When the time dimension is explicitly considered, a **dynamic** system or **dynamic** modeling results.

## Mapping Background Space

Measuring background space becomes important when we want to figure out the rules that arrange elements. In order to place elements into the background space, we need to be able to picture it and draw a map of it. If the background space can be drawn, a coordinate system can be defined.

In addition to the shape and size information described earlier, a map of the space requires that:

- A system of gridlines, much like a street map, is applied to the space.
- Once the streets are laid down, each space where an element can go is labeled.

A map of the background space is generally required to make precise models for testing with the scientific method.

# The Background Feature Extractor

**Container, Origin, Places**

The procedure to map the background space for a language-of-patterns analysis is given by the letters COP.

- Container
- Origin
- Places

We will use the filled square pattern structure as an example of the process.

**Container**

1. Imagine a container (sample boundary space) [**A**] to hold the system. The properties of the container should be the properties of the background space. Our structure is a set of square boxes with height and length arranged in a line. A two dimensional rectangle will contain the sample boundary space accurately.

- Draw a rectangle around the system.
- Since a rectangle has length and width, draw a line for each of these [**B**]. (In technical terms, this step defines the coordinate system of the space.)

define container

**Origin**   2.   Next we assign an origin point [O] in the container [C]. This is where we will start drawing the map. Notice that the container forms a boundary around the sample boundary space and the surrounding space.

assign origin

C = container
O = origin

**Places**   3.   Finally, we mark the potential background spaces by drawing background placemarker [D] spaces in the container.

  •   We draw them to a convenient size, in this case the size that will hold the squares. In this example the placemarker spaces are represented by:

4.   To find our way around we must label the placemarkers [E] just like naming streets on a map.

5. We remind ourselves that the container space is just a sample portion of the overall background space by extending the places marked with dotted lines [**F**].

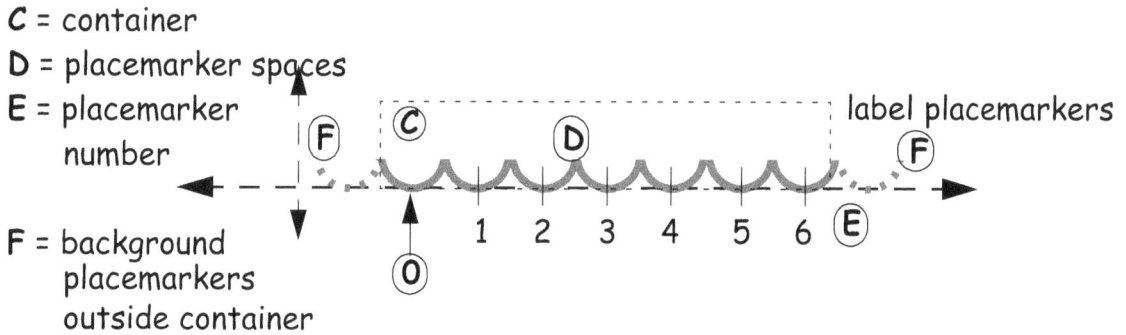

**C** = container

**D** = placemarker spaces

**E** = placemarker number

**F** = background placemarkers outside container

F  C  D  label placemarkers  F

1  2  3  4  5  6  E

O

6. With the **container** (and its coordinates), the **origin**, and the **places** established, the sample space is redrawn leaving out the imaginary container. The background space is now described and can be explored itself or can be used for the mapping of rules and the evolution of the system.

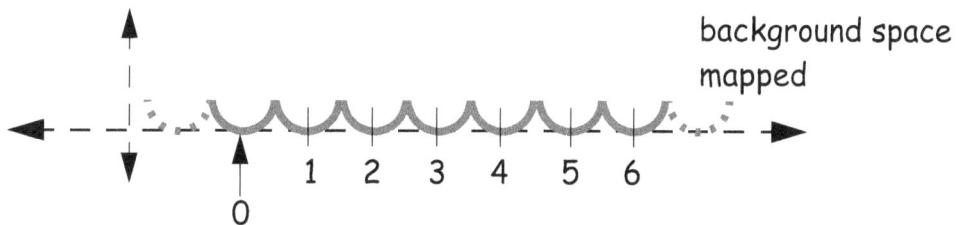

background space mapped

1  2  3  4  5  6

0

## Summary

An overall graphical summary of the background space mapping process is shown here:

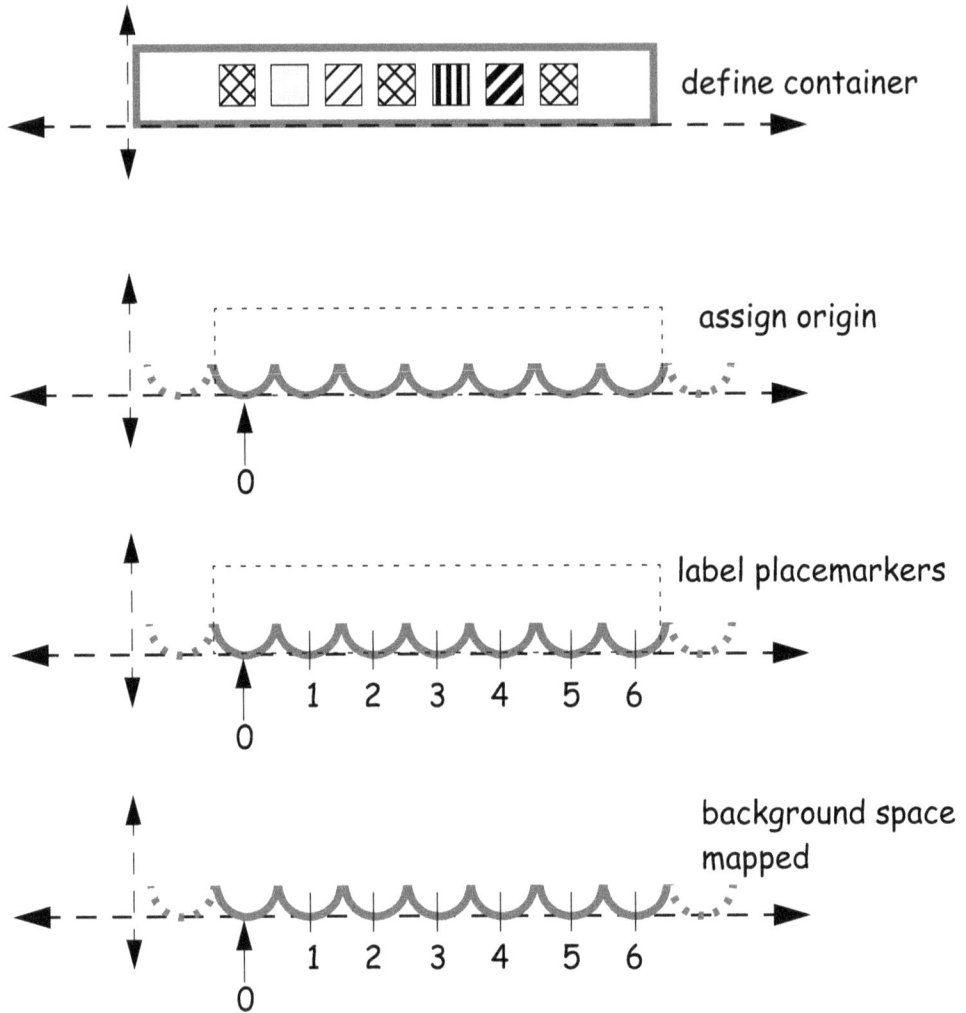

define container

assign origin

0

label placemarkers

1  2  3  4  5  6

0

background space mapped

1  2  3  4  5  6

0

Let's practice extracting the background space properties of some other systems.

- CAUTION: In this process, we frequently find assumptions that we have made to be incorrect. Always keep an open mind to the actual properties that you see.

**Practice Exercise** Extract the background space of this system. Remember that placemarkers can be a set of coordinates. (Answer on next page.)

system

container

origin

places

**Answer:**
**Container**

The first assumption is that this drawing exists in the flat plane of the page. Proceeding from that assumption, we draw a container and place the reference lines (axes) that make the container.

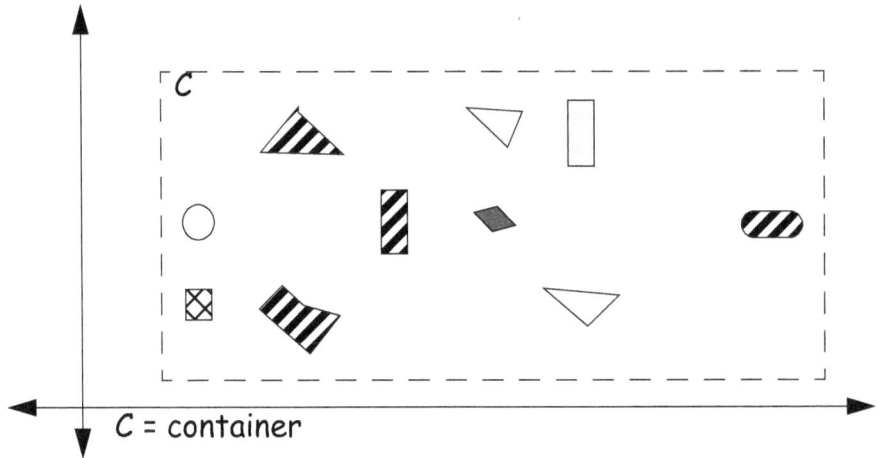

*C* = container

**Answer:**
**Origin**

Next we choose a convenient origin.

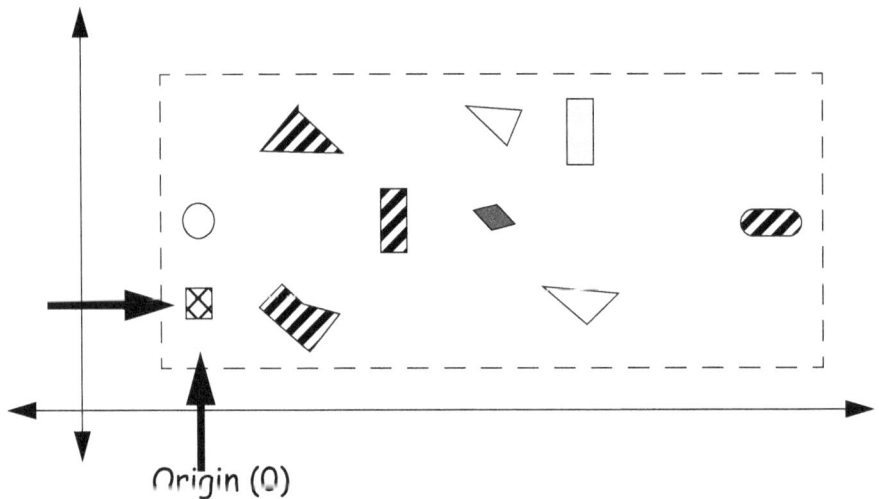

Origin (0)

**Answer: Places**

Finally, we draw in places to put things, placemarker spaces. Remember to extend the map with dotted lines into the background space beyond the sample boundary. [In solving this example, a Cartesian coordinate system in two dimensions has been drawn!]

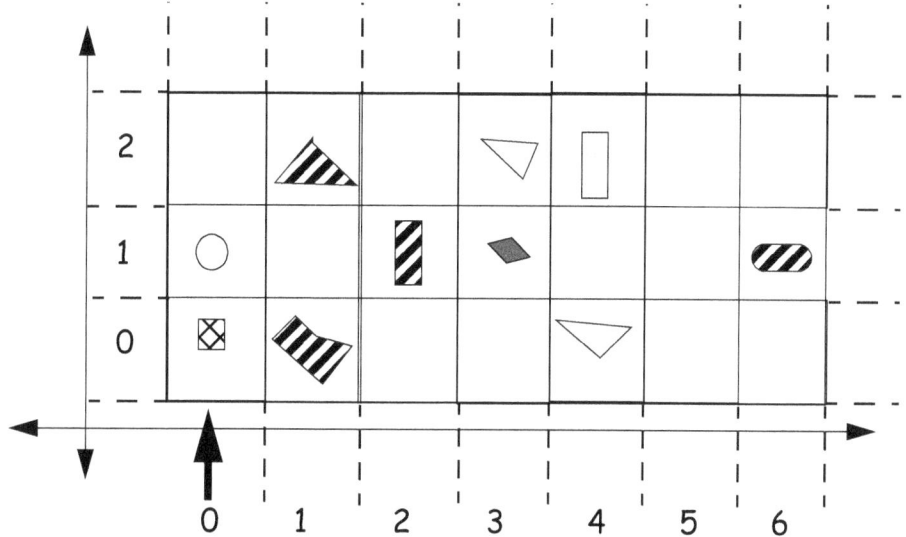

**Answer: Summary**

This description of background space is two dimensional in the plane of the page, due to the arrangement and properties of its elements (flat drawings). The location of each object can be identified by its coordinates.

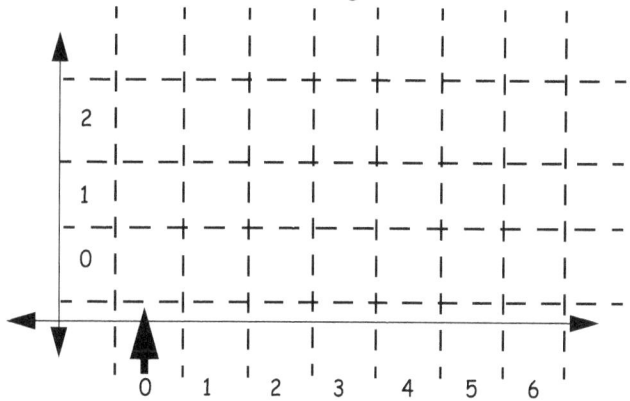

# CHAPTER 19

# Exploring Rules with the Graphical Analyzer System

| Analyzers | Application |
|---|---|
| Ordering Analyzer | Guide inferential process of extracting rules of arrangement by examining the ordering of elements |
| Rule Extractor | Extract the properties of background space in order to infer and describe rules in a system |
| | Define the coordinates and dimensions of a system |

## Arranging the Elements

The extraction of rules is an inferential process.   Elements are arranged in one of three ways in a background space.  They can be sorted based on an observable into each of these arrangements:

- Nominal arrangement

- Ordinal arrangement

- Spatial arrangement

### Nominal Sorting

Elements are always either <u>in</u> or <u>out</u> of the background space based on an observable.  This is called a **nominal** arrangement. Nominal arrangements are the same as grouping and sorting and use the Sorting Analyzer.

### Ordinal Sorting

Each element can be placed in a sequence based on an observable, such as size, age, weight, and time.   This is an **ordinal** arrangement. Ordinal arrangements are common and include food chains, sequences, and growth patterns. They require an **observable** and a **measurable** value associated with the observable that can be sequenced.

## Practice Exercise

Tadpoles have tails and frogs do not. Tadpoles lose their tails in proportion to their bodies as they turn into frogs. Propose an order for these tadpoles based on the observable of tail/body length. Answer on the next page.

ORDERING ANALYZER

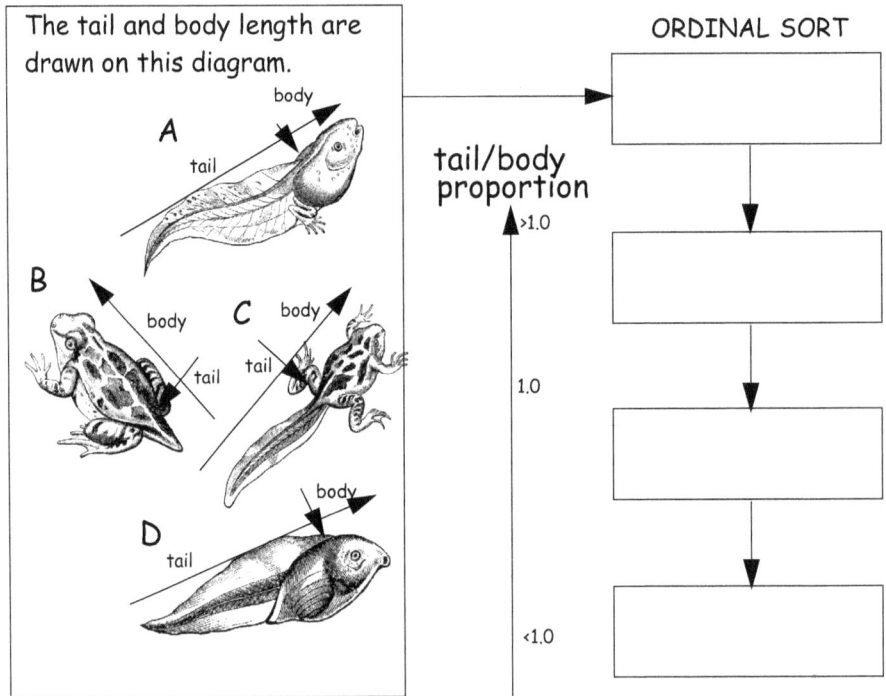

**Answer**   Answer to tadpole ordinal sort.

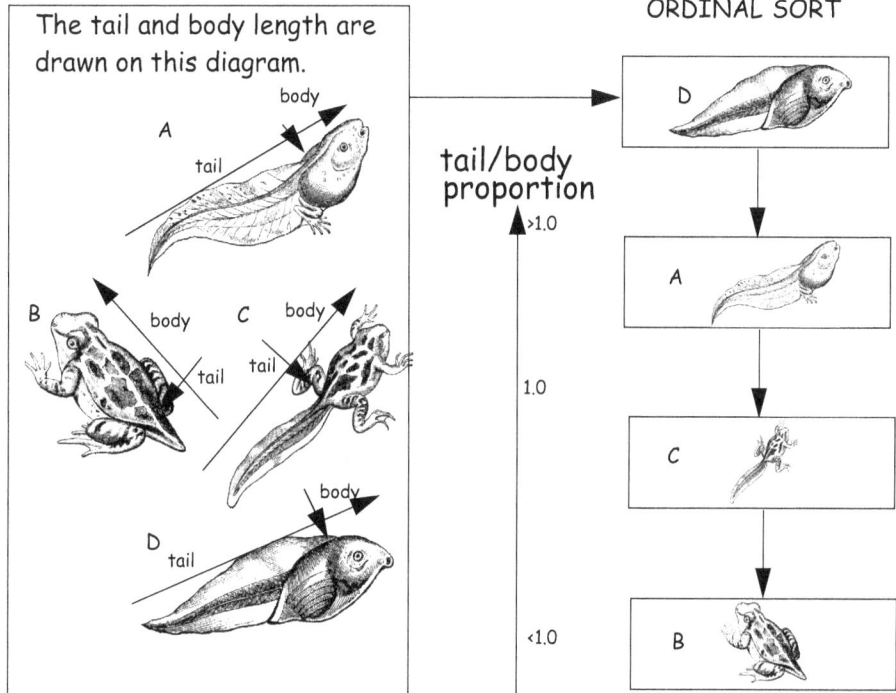

The tail and body length are drawn on this diagram.

ORDINAL SORT

tail/body proportion

## Spatial Ordering

Each element can be placed at a specific point in space based on the background map coordinates. This is a **spatial** arrangement. Ordering elements spatially is a physical mapping that places each point of an object in a space defined by a grid. There must be a measurement for each coordinate of the background space. Spatial ordering uses standard graphing techniques.

## The Rule Extractor

Inferring rules is a relatively advanced capacity that becomes available to students in middle school and develops in sophistication through the highest grades.

- A rule operates on an element.

- A rule is always defined in terms of the element on which it is operating.

· Rules arrange elements in some type of space.

· Rules either order elements with respect to some point in the background space, or they order elements relative to another element.

· Ordering elements with respect to a constant origin in the background space is easier to learn.

· We will only consider this easier case.

Rules are almost always inferred from the arrangement of the elements in a carefully defined background space. This is why the elements and the background are extracted first. Rules can be extracted by mapping a variable onto a coordinate system.

## Start, Move, Repeat (SMR)

The procedure to define the rules for a language-of-patterns analysis is given by the letters SMR.

· Start
· Move
· Repeat

To extract the rules we must always come prepared with:

· The structure
· The elements in the structure
· The extracted background space

We will use our familiar filled-square pattern as an example of the process.

The elements have already been extracted with the Element Analyzer.

ELEMENT ANALYZER

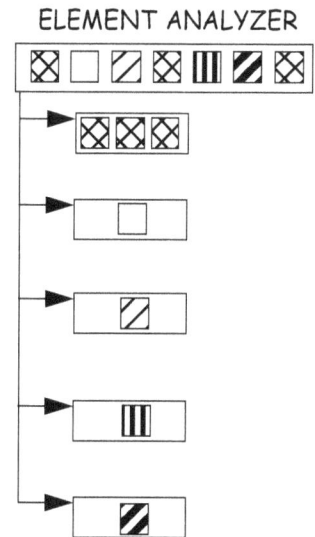

We have also already extracted and described the background space:

BACKGROUND SPACE EXTRACTOR

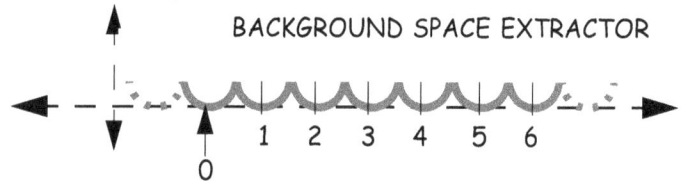

Now, we can extract the rules for the first element:

1.   Define where in the background the element **starts (S)**.

2.   Determine where the element **moves (M)** (what direction and how far).

3.   Determine how often the move is **repeated (R)**.

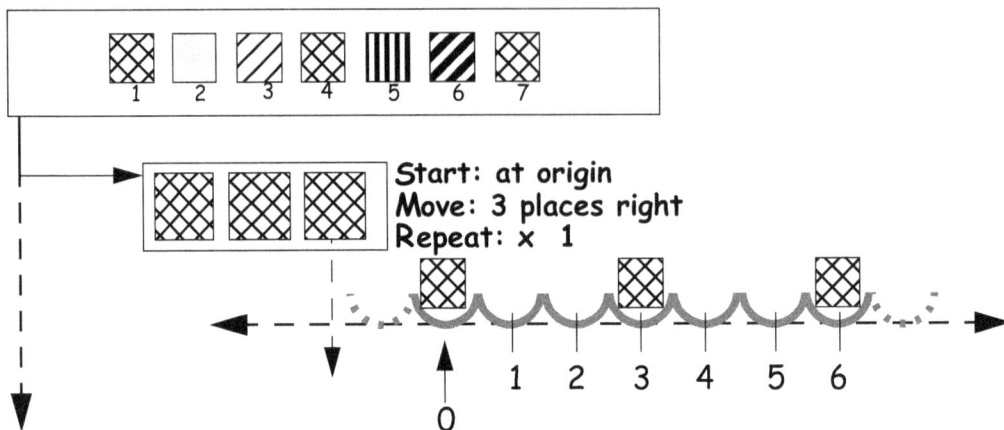

Now, we extract the rules for the second element:

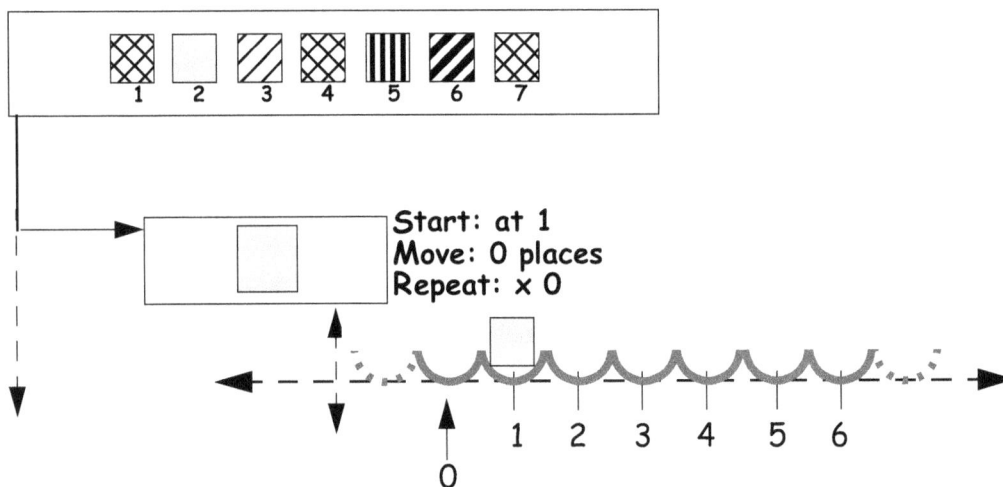

The rules arranging each element can be sequentially extracted in this fashion

**Practice Exercise** Complete the extraction process for elements 3, 4, and 5. Write the rules in the SMR space below. A summary of the rule extraction for this pattern of elements is shown on the next page.

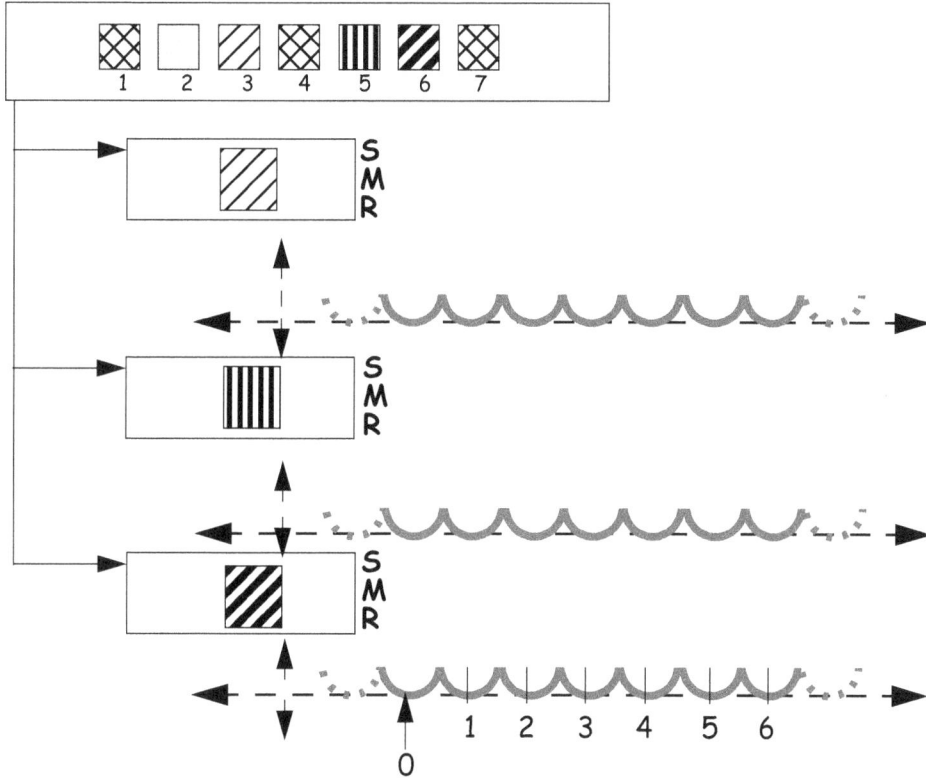

**Answer**   Completed rule extraction for the fill-pattern structure.

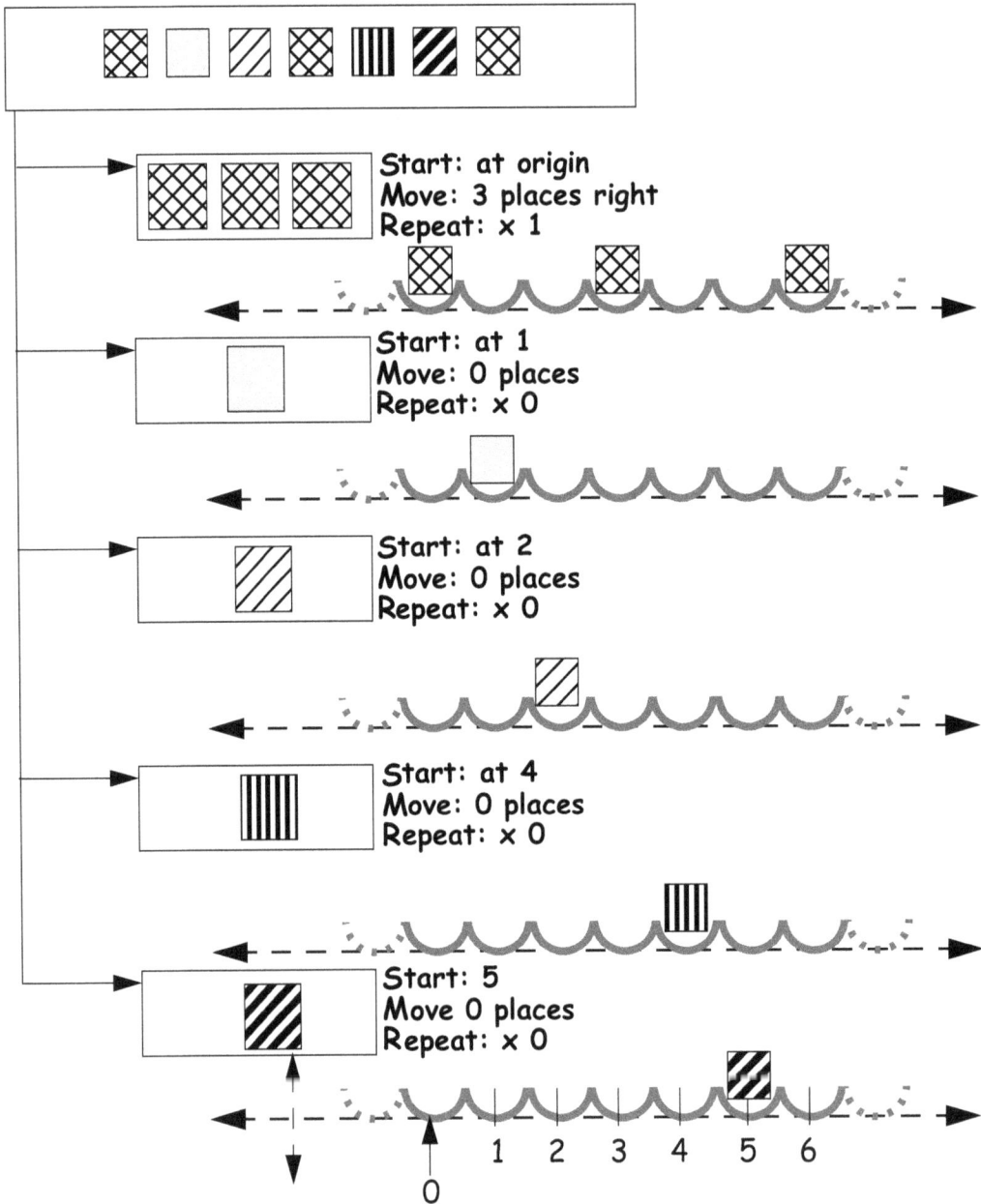

Start: at origin
Move: 3 places right
Repeat: × 1

Start: at 1
Move: 0 places
Repeat: × 0

Start: at 2
Move: 0 places
Repeat: × 0

Start: at 4
Move: 0 places
Repeat: × 0

Start: 5
Move 0 places
Repeat: × 0

0   1   2   3   4   5   6

This Rule Analyzer is very simple to use in terms of step-by-step processes, but it can get complicated as the structures grow in size, element categories, and dimension. Remarkably, this process is sufficient to extract the rules and write them clearly and precisely for the majority of the systems likely to be encountered in the natural world.

For example:

- In biochemistry and medicine, the determination of the shapes and functions of proteins and enzymes are determined by characterizing the arrangements of atoms extracted from the pattern of x-ray crystallography.

- The complex and complicated spectral fingerprints of stars and chemical compounds are described and interpreted by understanding the rules that arrange their observables of light, color, and chemical reaction.

## Patterns and Rule Extraction

Generally, when we talk about patterns, we are referring to a certain type of structure in which certain elements are following a repeating rule of arrangement.

- Finding the pattern in a structure is a rule extraction.

- This is why describing a pattern in terms such as ABABAB defines the kind of repetition that an element will have, but it does not describe the overall composition of the structure.

- The Language of Patterns can describe the repeating rules, elements, and backgrounds of a typical pattern, as well as efficiently describe any system of elements or structures.

**Practice Exercise** This example **is** rocket science. Extract the rules of this rocket's motion. Here is a hint: the motion of a rocket is an arrangement of positions of that rocket.

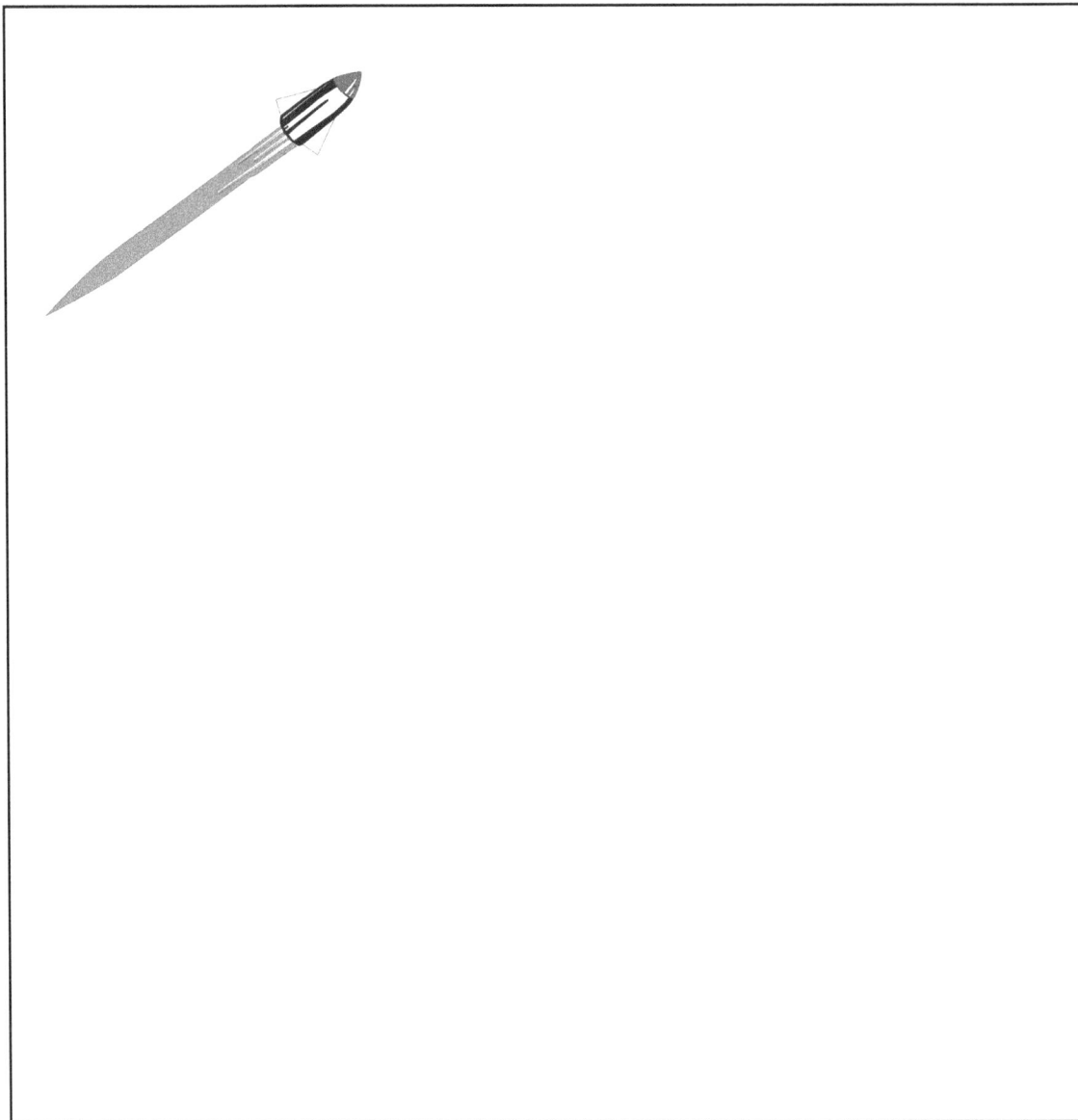

**Answer**  The rocket's motion is described by the sequential arrangement of its position as it moves through space.

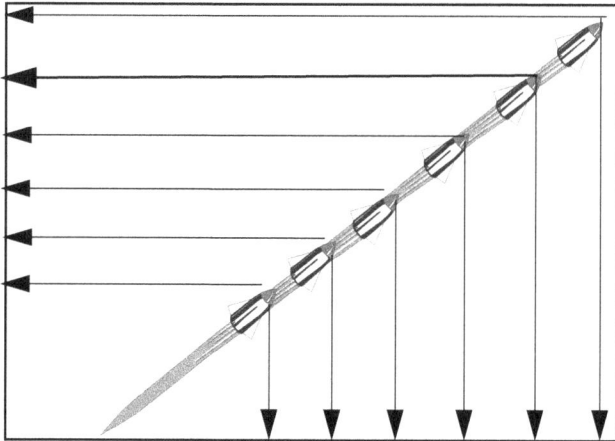

The elements of the motion are six points in a horizontal direction and six points in a vertical direction

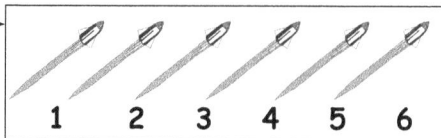

1  2  3  4  5  6

The background space is a square container with a vertical (y) and horizontal (x) axis.

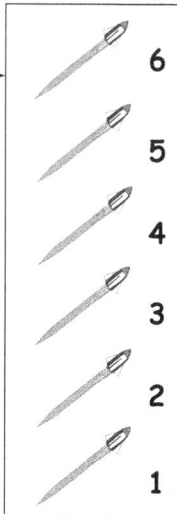

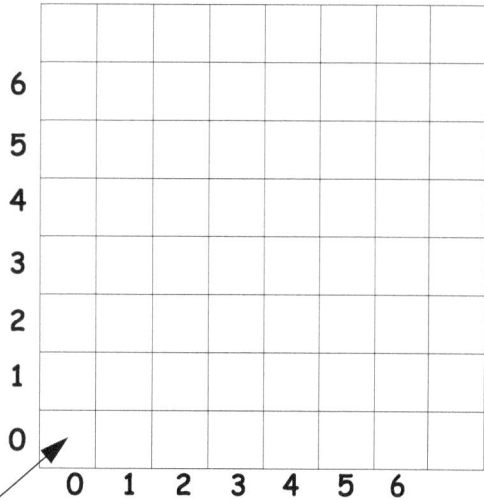

6
5
4
3
2
1

6
5
4
3
2
1
0

0  1  2  3  4  5  6

Origin

The origin is chosen and the placeholders are a grid on the x and y axes

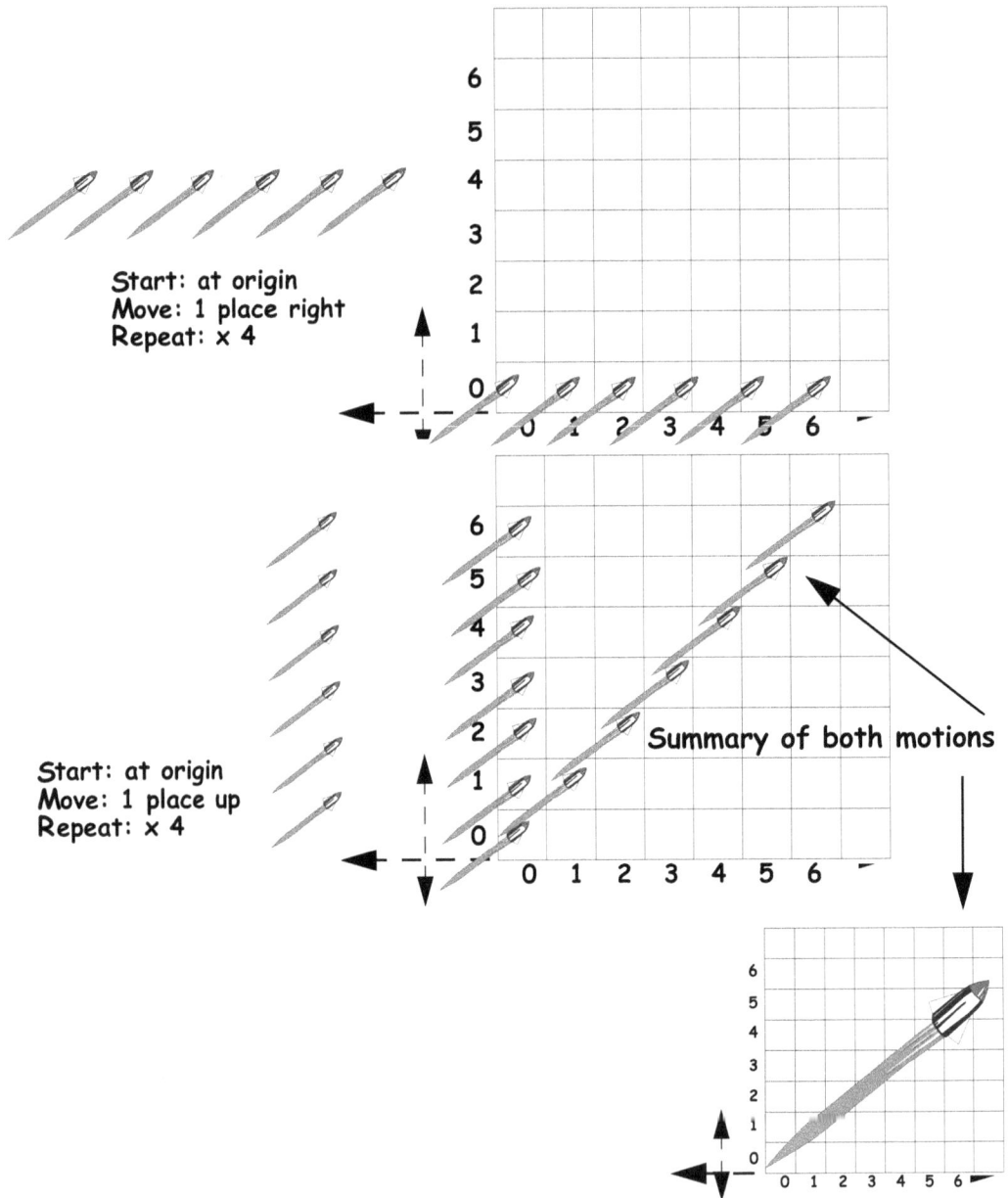

Start: at origin
Move: 1 place right
Repeat: x 4

Start: at origin
Move: 1 place up
Repeat: x 4

Summary of both motions

# CHAPTER 20

# Using the System and Structure Graphical Analyzers

## Extending Systems Description

| Analyzers | Application |
|---|---|
| System Analyzer<br>Structure Analyzer | Characterize the composition of a system or structure in terms of its elements, rules and background space |

## Describing a Structure

Describing a system or a structure requires knowledge of the elements, rules, and background, as well as how they go together.

The separation of a system into its parts is a concrete activity that can be performed by students as early as first grade. As the student's skills grow and the learner matures cognitively, the same analyzer tools can be used for more sophisticated exploration. This includes:

- The contrast and correspondence between the organization of systems.

- The inferred knowledge of how a system is organized.

- The ability to use one system as a metaphor or model for another.

## Defining Systems

When elements are arranged in a particular background space by a set of rules that describe the relationships among the elements, background itself, and other rules, we have a **system**. Systems give rise to emergent properties and can be arranged in a larger space by rules that apply to a level of systems such that they can be arranged to form larger systems. So at one level, systems can be treated like elements in a larger system, that is, systems are formed from sub-systems arranged in space.

The power of the Language of Patterns is the use of the same critical process to find connections in systems of objects and ideas. The same process that allowed us to describe how things go together can be used again and again until we see all the parts and relationships both at a smaller and at a larger level.

**Example** Consider the systemic structure of a very familiar game such as kickball.

· Make a list of what is needed to play kickball.

- First, we need to get the players together. Ideally, we need 12 -18 people.
- We need a kickball.
- We need four bases.
- We need a place to play.
- We are going to have to pick two teams from our players.
- When a team takes the field, it will have a pitcher, a first, second, and third base player. Each team needs a catcher and at least one or two outfielders.
- Another element of the game will be the number of runs gained by each team as they play. This will be kept as a tally called the score.

| Kickball list | |
|---|---|
| Billy | Suzie |
| Linda | Sam |
| Joshua | Peter |
| Cindy | Catherine |
| Chris | Tom |
| Barbara | Earl |
| | |
| four bases | |
| kickball | |
| field on Vine Street | |
| catcher | |
| second base player | |
| first base player | |
| third base player | |
| pitcher | |
| score | |

- The structure of a kickball game can be drawn graphically on the Structure Analyzer:

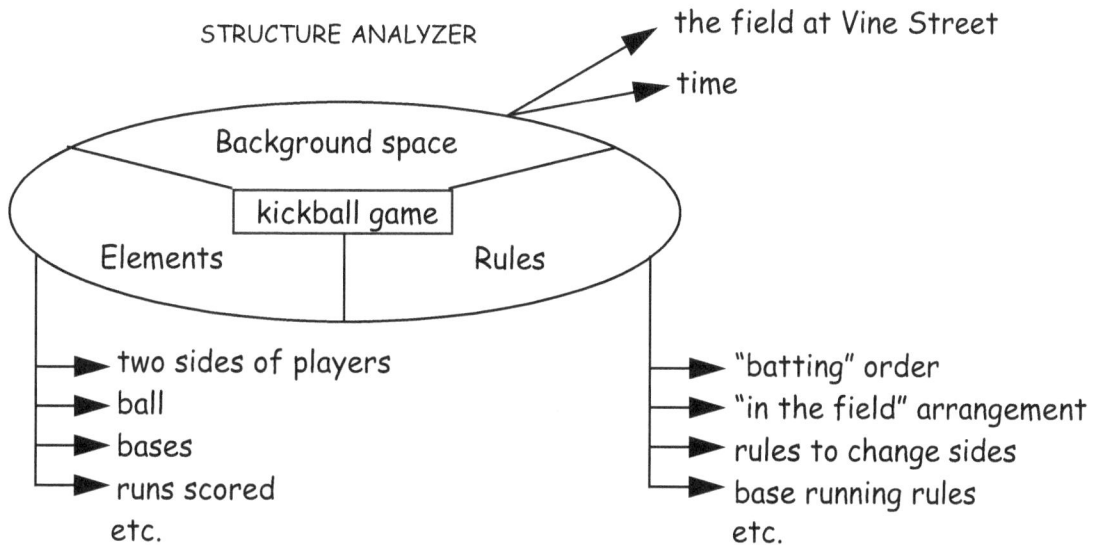

STRUCTURE ANALYZER

the field at Vine Street

time

Background space

kickball game

Elements

Rules

- two sides of players
- ball
- bases
- runs scored

etc.

- "batting" order
- "in the field" arrangement
- rules to change sides
- base running rules

etc.

- This kickball game has a structure that includes:

    - A background space - a place and time to play.
    - Elements - two teams or sides, a ball, bases, and a scoring system.
    - Rules - arranging all of the different elements into batting sides, fielding sides, ways to change sides, rules for keeping score and then determining who wins, etc.

Each of the components of the structure (elements, rules, and background) have their own structure which can be described by elements, rules, and background.

Let's consider just the ball.

- The ball is an essential element of the game.
- It has certain properties that define it as a kickball.
- These properties are the result of its structure.

Right now, lets just be concerned with the properties that describe the kickball. Either the Property Analyzer or the Element Analyzer can be used for this purpose.

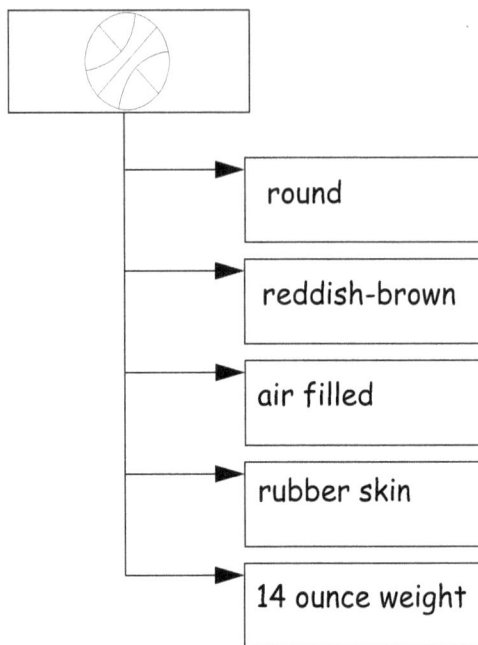

round

reddish-brown

air filled

rubber skin

14 ounce weight

ELEMENT ANALYZER

Element analyzers can chart properties to whatever degree of precision is needed for the topic of interest.

Most of us may not need to know or care about the actual measurement, materials, or air pressure need to make the kickball regulation, but they are properties of the ball none-the-less.

The background space is very important for the structure of the game. Just a couple of examples can serve to highlight this. First, we must consider the properties of the background space.

- Is the kickball game to be played after school on a night with a lot of homework?

    - Then the game will be limited in terms of its length, because the player elements will have to go home to do their homework (if they are allowed to play at all!).

    - A Saturday game might have no limit on time, but the sun might be in the eyes of the kicking team in the morning hours, leading to very different games on the weekend versus after school.

    - The same rules and players, and elements are all affected by the background space or context of the game, in this case the time of the game.

- What if the game is played on a field with woods behind the outfield versus a city sand-lot with a brick wall at the back of the outfield?

    - A kick out of the outfield and into the woods will surely be a home run.

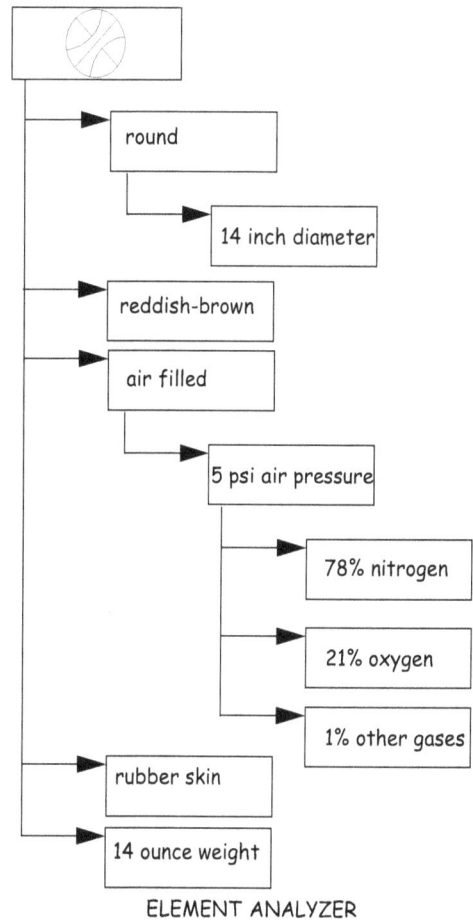

round

14 inch diameter

reddish-brown

air filled

5 psi air pressure

78% nitrogen

21% oxygen

1% other gases

rubber skin

14 ounce weight

ELEMENT ANALYZER

- However, a kick out of the outfield into the brick wall can be caught on the fly and may be an out!
- The same rules and the same players again are affected by the background space.

- Finally, the background space determines what is in-bounds and what is out-of-bounds. If the ball falls inside the boundary line, one set of rules apply to the players and ball. If the ball falls outside the boundary line, play proceeds differently.

Background space and how it is structured have a huge impact on the way the game itself will be structured.

# CHAPTER 21

# Evolution and Change with Graphical Analyzers

| Analyzers | Application |
|---|---|
| Evolution and Change Analyzer | Identify and describe the part of a system that causes it to rearrange and exhibit changed properties |

## More on Systems and Structures

We have been talking about **systems**, **structures** and **elements** throughout this Tour.

- Usually, we think of an element as being a simple object, assembled into a larger and more complicated structure. We might think of a nail as being an element of a table. The relatively complicated arrangement of wood, screws, nails, and brackets is usually thought of as a structure. However, if the table is viewed as part of a much larger system, a cafeteria for example, it becomes an element of the system of tables, chairs, food, waiters, and customers.

- Alternatively, the chemical and physical organization of the nail itself is a highly complicated arrangement of smaller chemical elements. The arrangement of these smaller objects constitutes the structure of the nail.

- The following example helps us to appreciate that each element of a system has its own internal structure and its own properties and is therefore a (sub)system.

- In general, elements, rules, and background spaces each have properties and can be analyzed as if they were (sub)systems themselves. Analysis of any of these can be done by the Language of Patterns. The essential choice of the observer is to decide how closely the system or object needs to be examined.

- Systems can change when some force changes their elements, rules, or background space. A new system is the result. A method of mapping these changes is our concern now.

## Evolution or Change Analyzer

Structures and systems undergo change. When they change, they typically develop new properties. These new properties are the result of a specific change to some aspect of the original system. The process of change must be described in terms of the alteration of the original elements, arrangements, or background space. Often systems change over time. As was mentioned earlier change with respect to time is called dynamic change.

For example, if we <u>changed the space</u> upon which our now familiar series of squares lays from a within a rectangle to a donut, <u>the system and its structure will change</u>.

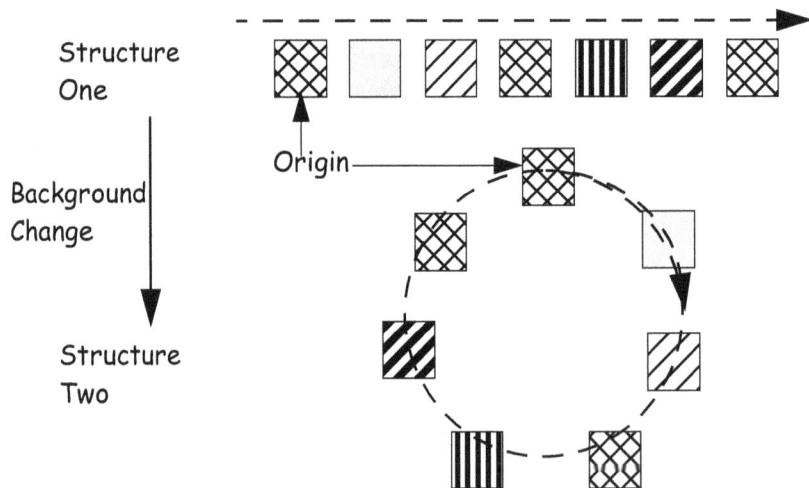

**FIGURE 60.**

The elements are the same (seven textured squares). The rules are the same (identical order and arranged along a central line through the center of each square). Most of the properties of the background space are the same (everything is still in the plane of the page). Yet by changing the shape of the background space (a circular rather than a linear space), the structure and the properties of the system are very different.

The Structure Analyzer can be used to describe the linear structure:

STRUCTURE ANALYZER

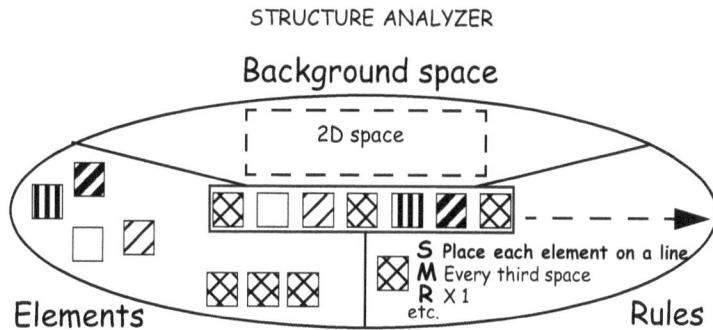

**FIGURE 61.**

We can describe the change in structure from a linear structure into a circular one by using a *Change and Evolution Analyzer*.

Change (or evolution) occurs when some force acts on one of the elements, rules, or background of one structure to make it change into another structure.

CHANGE AND EVOLUTION ANALYZER

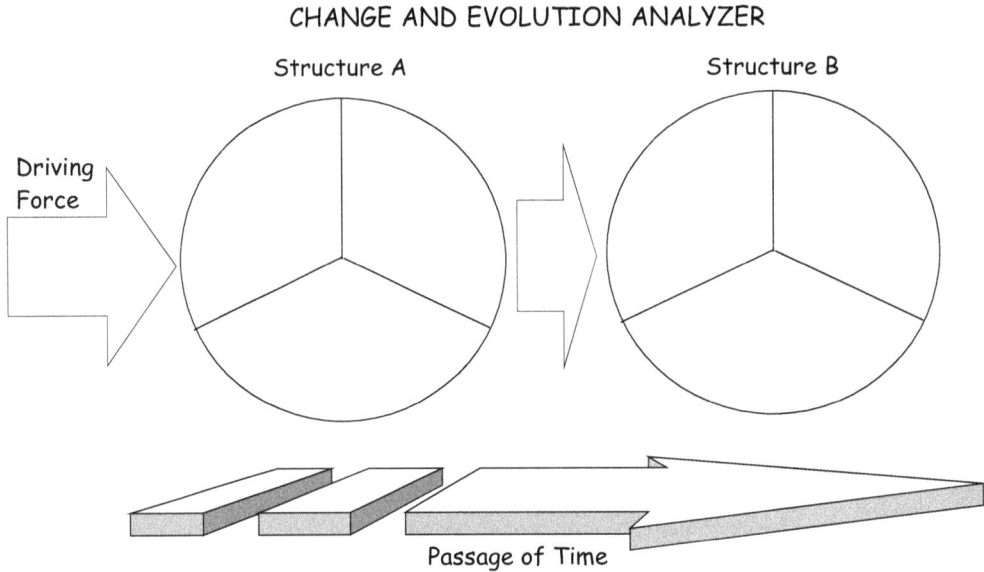

Structure A

Structure B

Driving Force

Passage of Time

In the case of the evolution of our "fill-pattern", a force acted to change the surface for "unfolding" the elements. Instead of having the order of the elements unfold on a linear surface, the surface changed, and the elements unfolded onto a curved surface:

EVOLUTION AND CHANGE ANALYZER

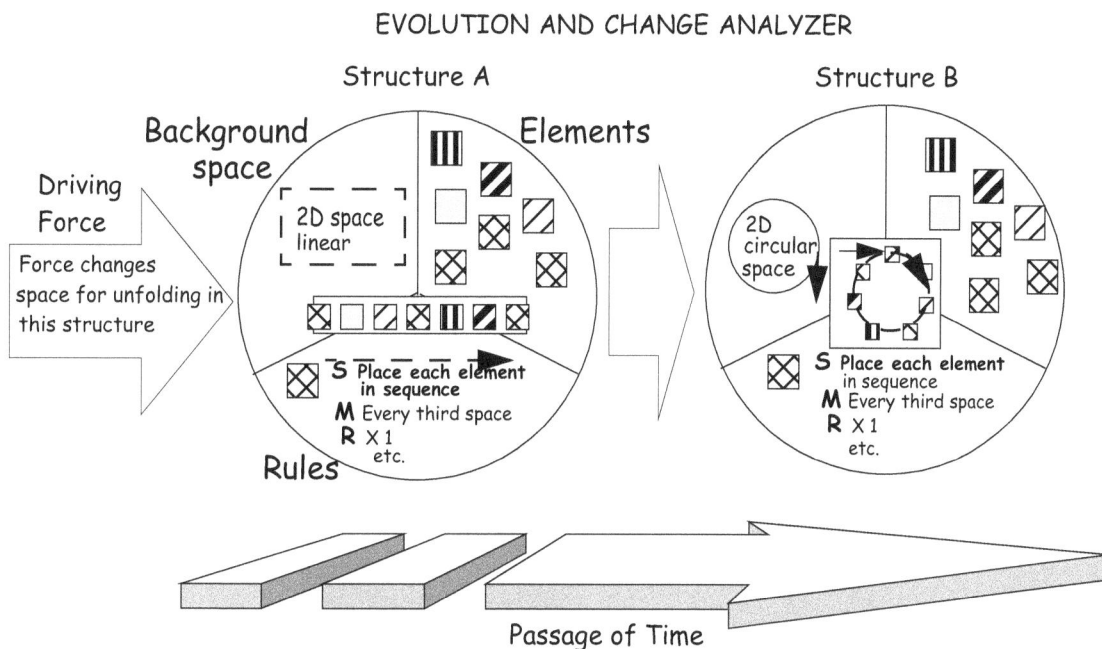

Both structures are similarly ordered and share many of the same properties, but clearly they are different structures with different functions.

**Example**   What happens to a stable pond ecosystem when a drought occurs?

First, the structure of the pond must be described. This is done using a Structure Analyzer. The force acting on this pond is a severe drought in which the region is over 20 inches behind in rainfall (the conditions in the Northeastern USA in 1999).

Where does this force have its effect? The background space is not changed nor are the rules. However, the drought reduces the water table and the amount of water entering the pond by rainfall. Therefore, the force is directed at the element of water, which is reduced in the system.

The Change Analyzer will change over time, determined by the relationship of water to the arrangements between the living and non-living elements. If we suppose that decreased water will reduce the survivability of the algae and bacteria in the pond, then we might expect the food chains to fail to support the protists, fish, and insects that eat the plants. The plants might overgrow the pond (they will probably have enough water) and lead it toward a marsh, which then might become filled in first with wetland ferns and rushes, then locusts, beeches, birches, and finally, over time, maples and oaks.

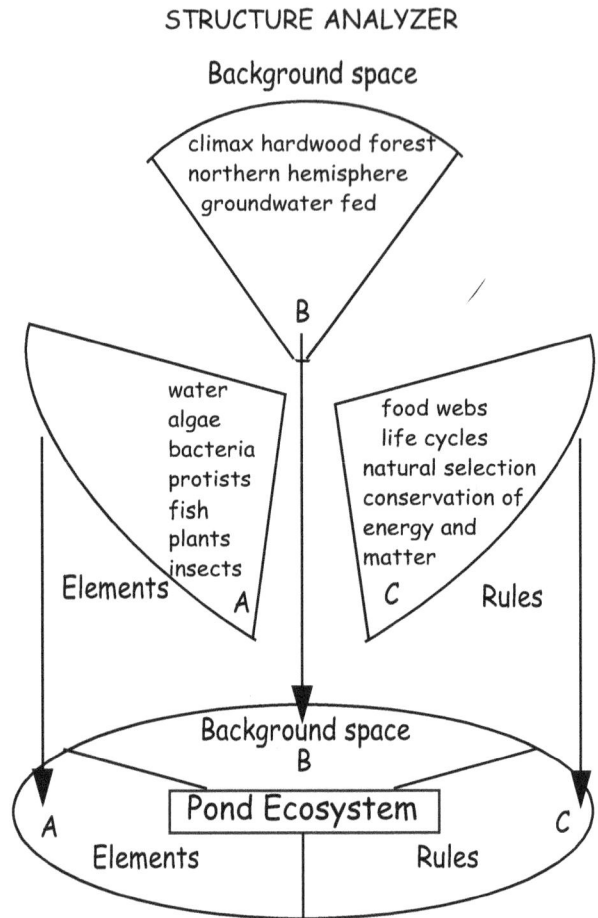

STRUCTURE ANALYZER

Background space

climax hardwood forest
northern hemisphere
groundwater fed

B

water
algae
bacteria
protists
fish
plants
insects

Elements

A

food webs
life cycles
natural selection
conservation of
energy and
matter

C

Rules

Background space
B

Pond Ecosystem

A

Elements

Rules

C

The Change Analyzer might show these multiple stages as the pond evolves out of existence and into a portion of wetland climax forest, all due to the drought of 1999.

## EVOLUTION AND CHANGE ANALYZERS

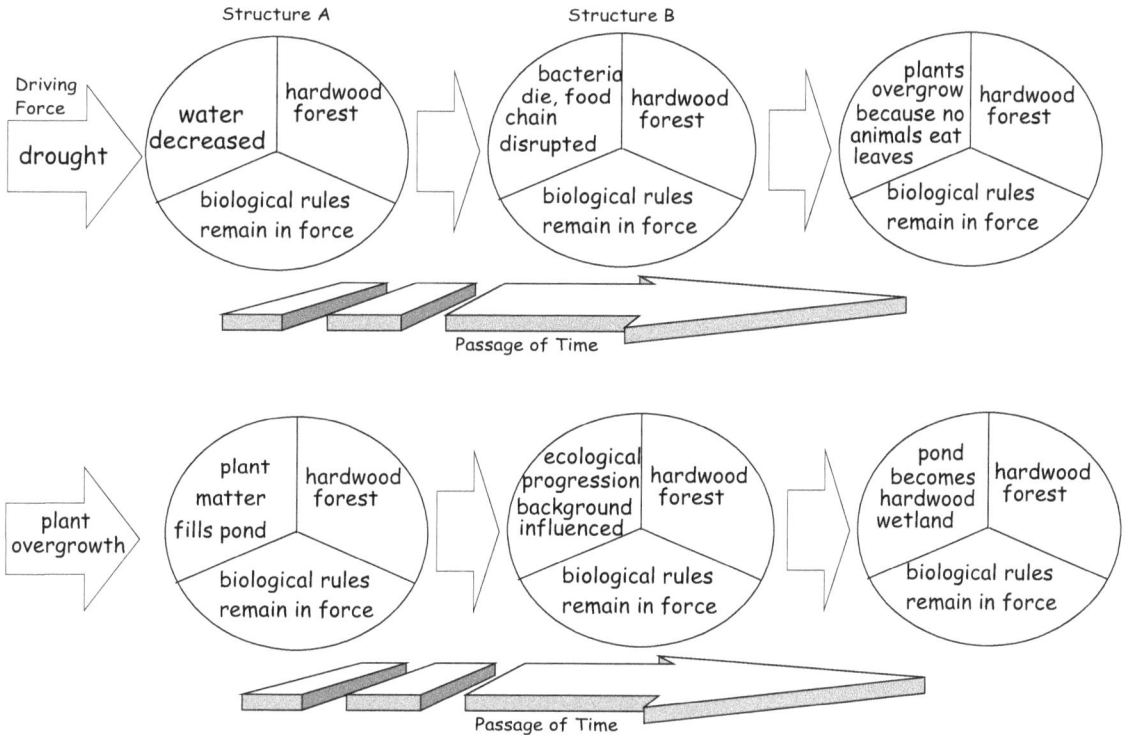

### Cross-Curricular Example: A Story Map

In this example, the Change and Evolution Analyzer acts as a standard Story Map, using a traditional reading/writing school exercise. The process not only extracts the story but allows a complete critical analysis to be immediately composed.

## CHANGE AND EVOLUTION ANALYZER

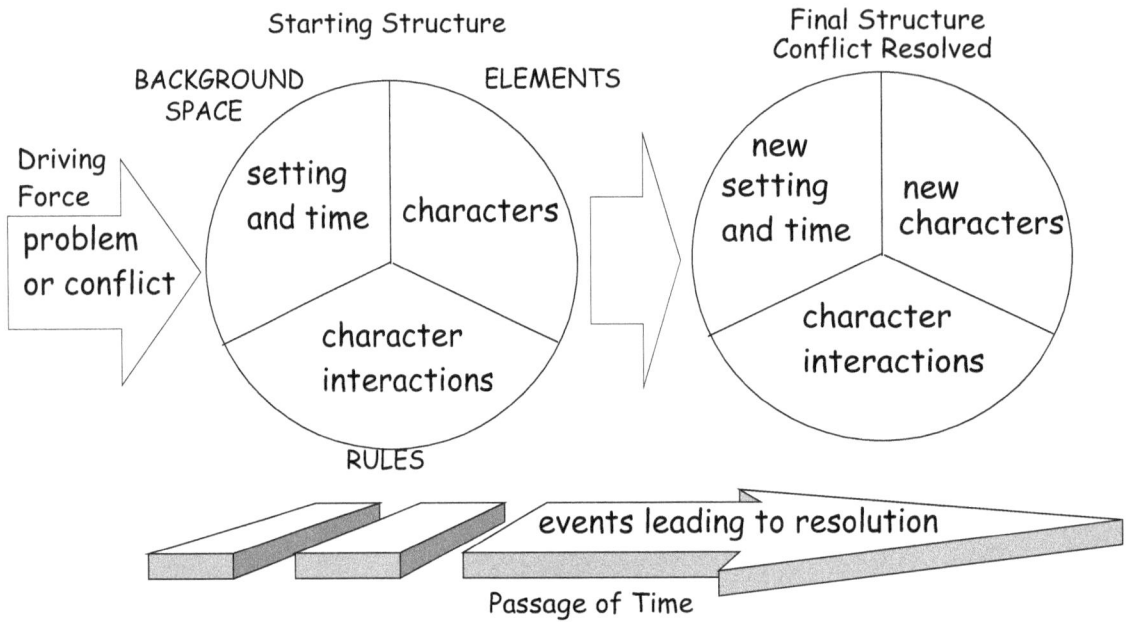

Starting Structure

Final Structure
Conflict Resolved

BACKGROUND
SPACE

ELEMENTS

Driving
Force
problem
or conflict

setting
and time

characters

new
setting
and time

new
characters

character
interactions

character
interactions

RULES

events leading to resolution

Passage of Time

# CHAPTER 22

# Analysis of More Complex Systems

| Analyzers | Application |
|---|---|
| Compare and Contrast Analyzer | Structures analysis of comparison between systems |
| Conflict Analyzer | Defines conflicts between two observers of the same system |

## Compare and Contrast Analyzer

An attribute Compare and Contrast Analysis is an observable/feature by observable/feature sorting of two systems into a Sorting Analyzer. There are two (or more) independent systems being sorted by **the same observable** in a Compare and Contrast Analyzer. The operation of comparing and contrasting attributes is best viewed as a pair of side-by-side sorting keys with a shared bin into which are placed the common observable properties. The process is a sequential one in which an element from the sorting bin on the left is compared sequentially to each element from the sorting bin on the right.

COMPARE AND CONTRACT ANALYZER

| elements to be sorted | elements to be sorted |
|---|---|

some observable

| does not have observable property | has observable property | has observable property | does not have observable property |
|---|---|---|---|

| contrasting properties | shared properties | contrasting properties |
|---|---|---|

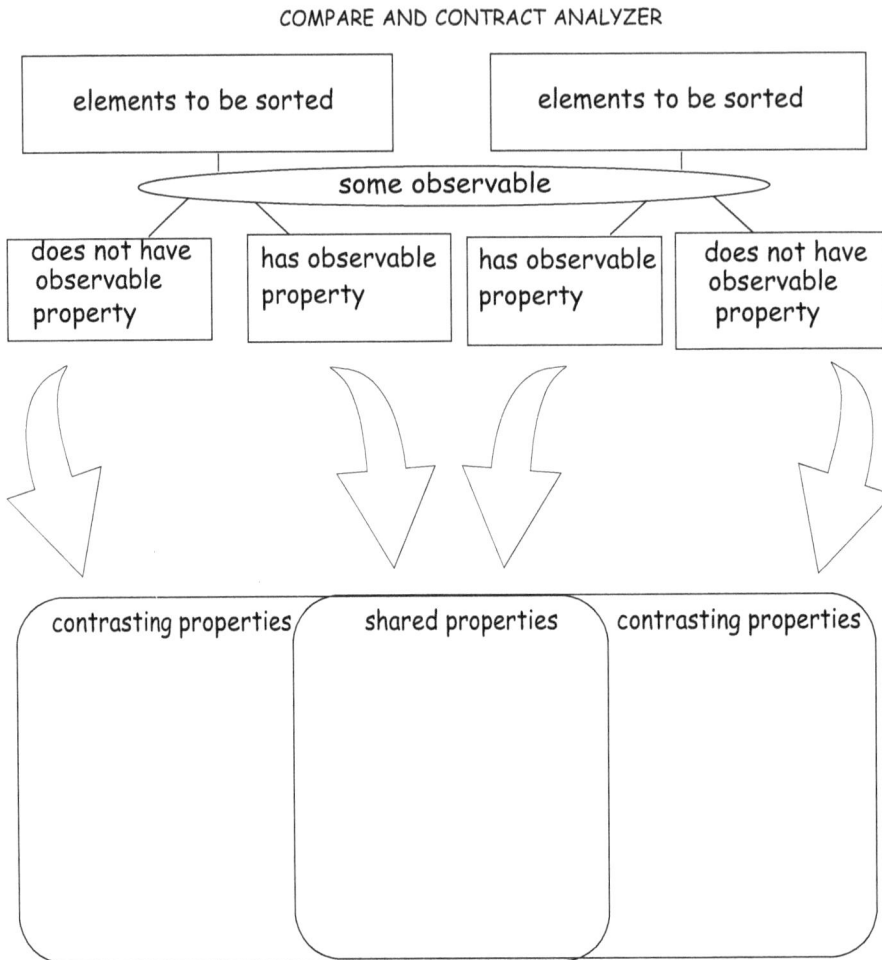

The process of comparing features in this way allows comparison, categorization, and identification of the differences and similarities between systems. This is an important step to finding connections and relationships that can be modeled and tested.

Other Compare and Contrast Analyzers are used in the Graphical Analyzer System. Often an important test is to compare the separate properties of two systems and observe if they are the "same". If all of the properties are the same, then the two systems are judged to be equivalent.

**Practice Exercise** Compare and contrast the elements in these systems of geometric figures (answer next page).

COMPARE AND CONTRAST ANALYZER

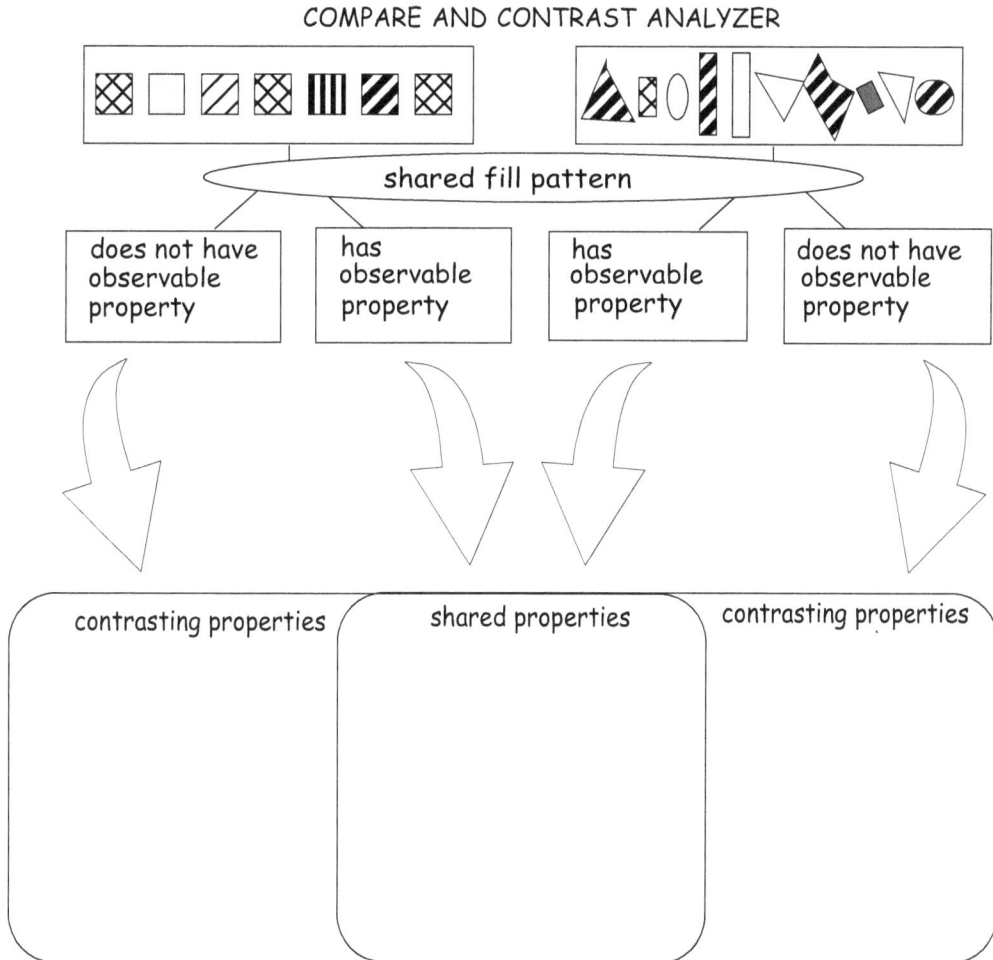

**Answer**  Answer to previous page.

COMPARE AND CONTRAST ANALYZER

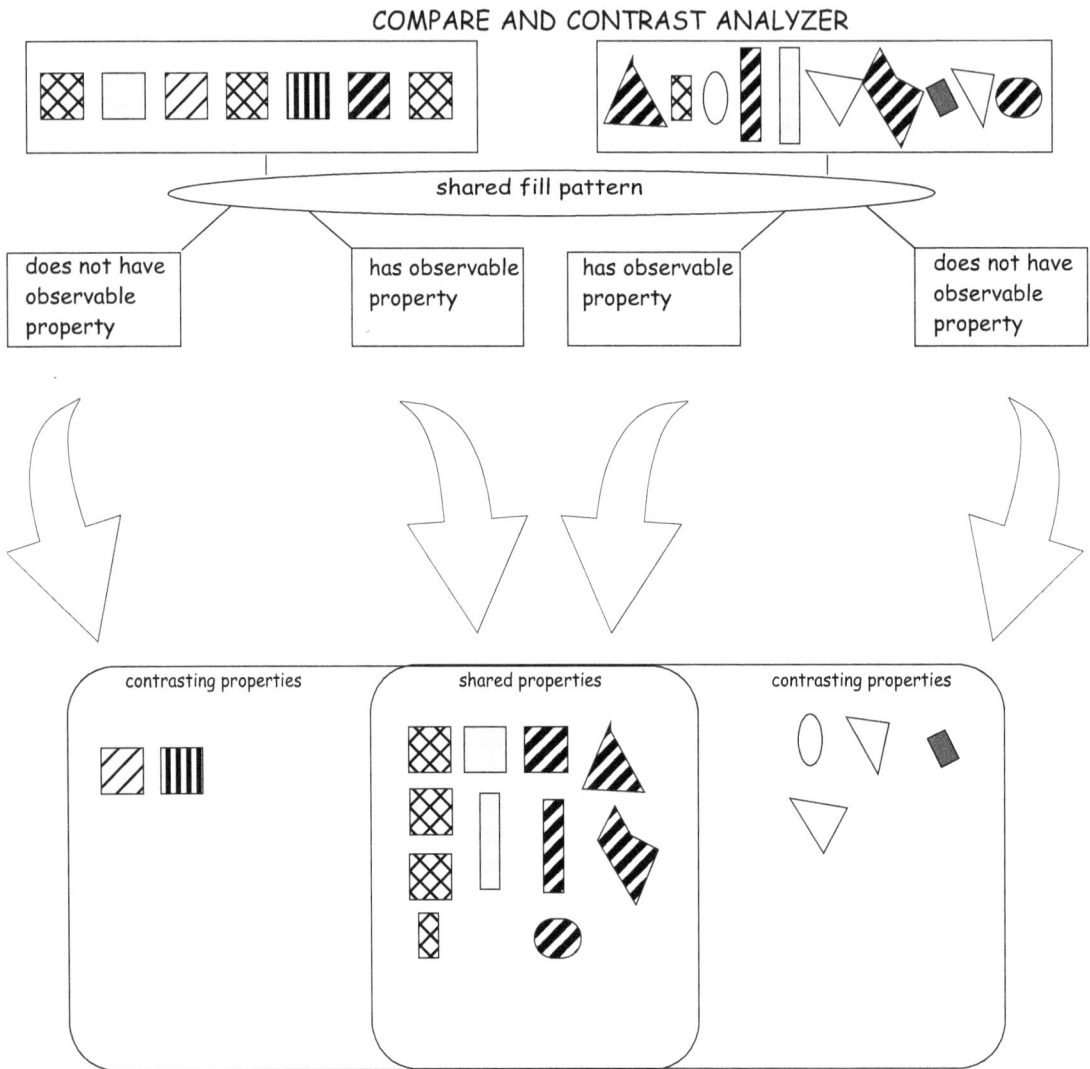

shared fill pattern

does not have observable property

has observable property

has observable property

does not have observable property

contrasting properties

shared properties

contrasting properties

**Practice Exercise** Compare and contrast the properties associated with electrical and magnetic forces. The mental process tests each property on the left with each on the right in sequence.

COMPARE AND CONTRAST ANALYZER

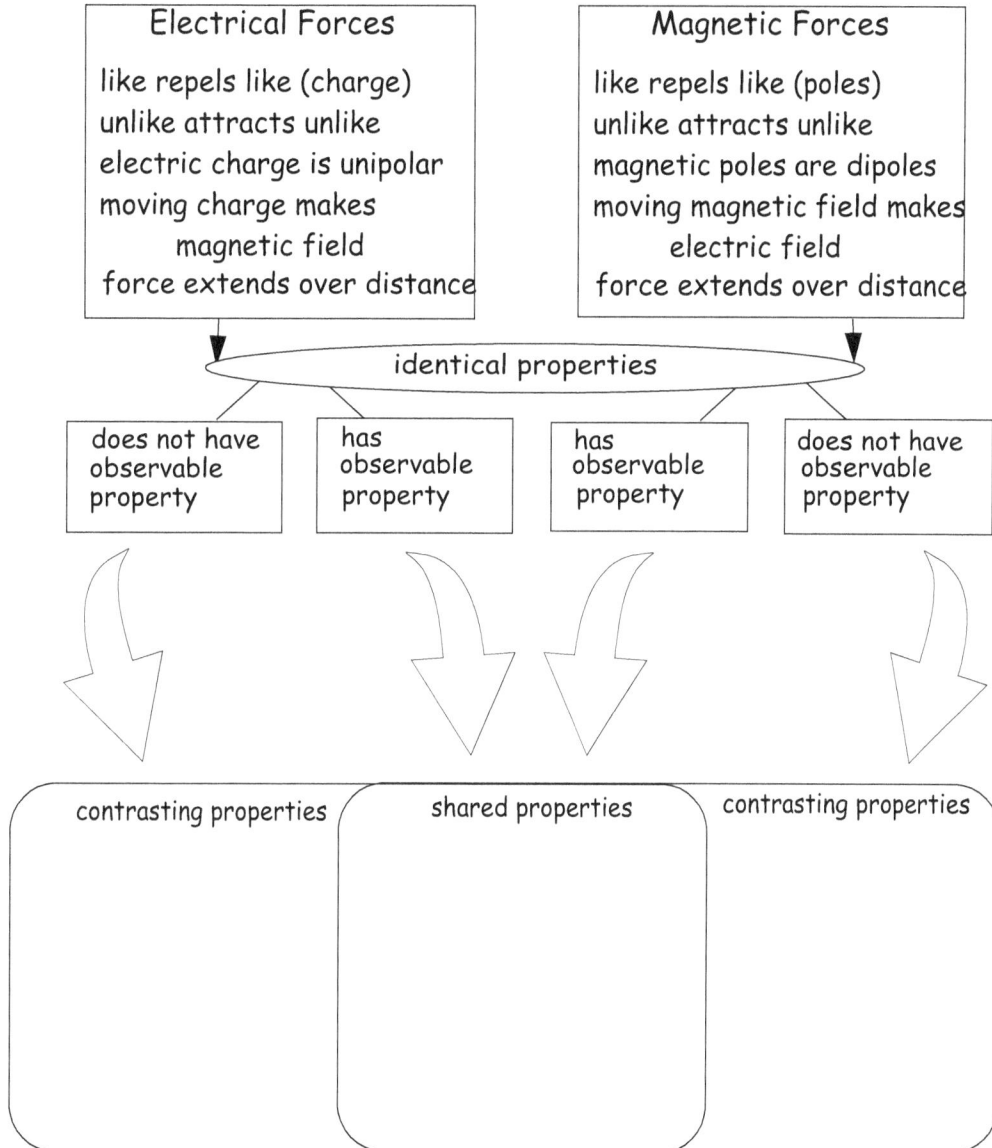

| Electrical Forces | Magnetic Forces |
|---|---|
| like repels like (charge) | like repels like (poles) |
| unlike attracts unlike | unlike attracts unlike |
| electric charge is unipolar | magnetic poles are dipoles |
| moving charge makes magnetic field | moving magnetic field makes electric field |
| force extends over distance | force extends over distance |

identical properties

| does not have observable property | has observable property | has observable property | does not have observable property |

| contrasting properties | shared properties | contrasting properties |
|---|---|---|
|  |  |  |

**Answer**  Answer to magnetic and electric force comparison.

COMPARE AND CONTRAST ANALYZER

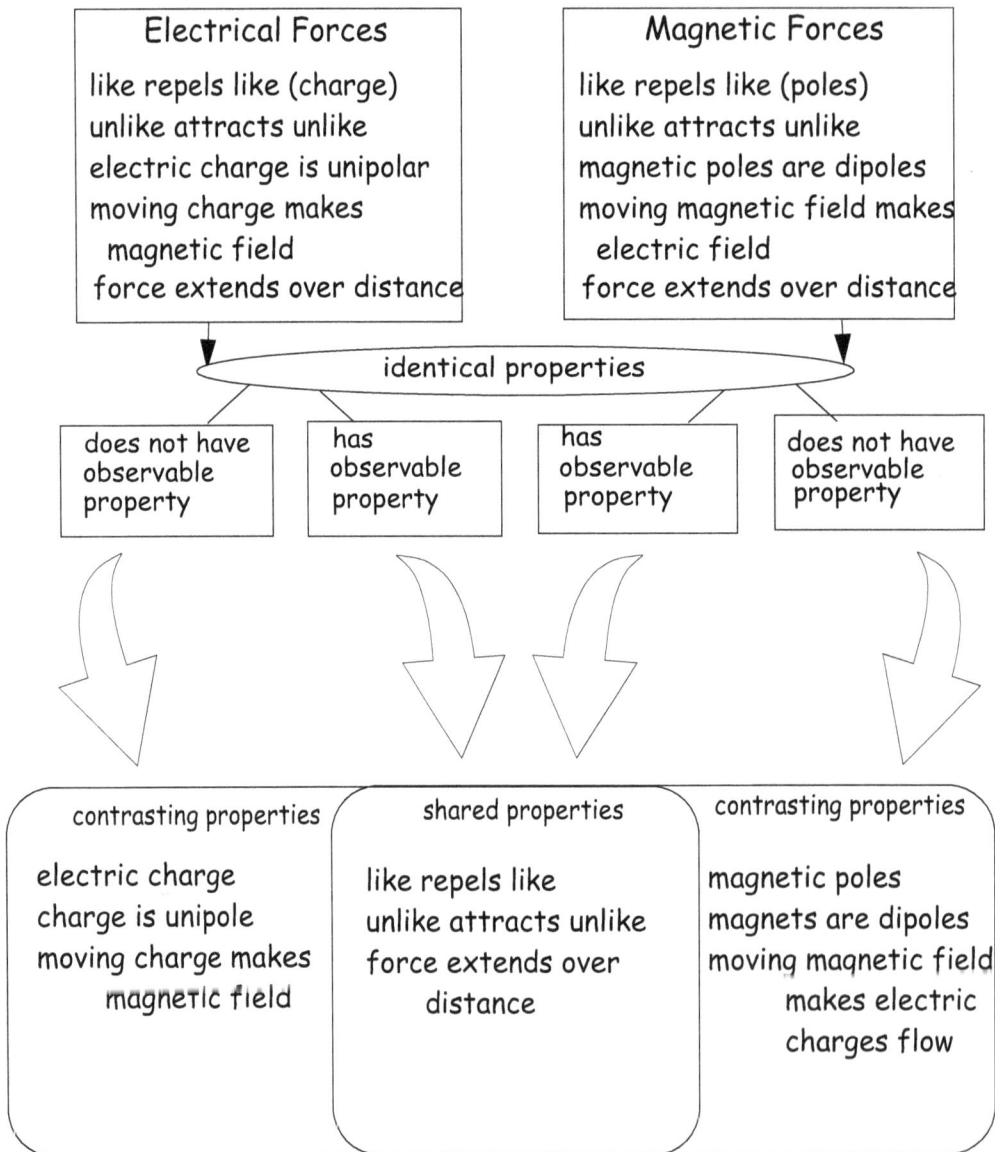

| Electrical Forces | Magnetic Forces |
|---|---|
| like repels like (charge) | like repels like (poles) |
| unlike attracts unlike | unlike attracts unlike |
| electric charge is unipolar | magnetic poles are dipoles |
| moving charge makes magnetic field | moving magnetic field makes electric field |
| force extends over distance | force extends over distance |

identical properties

| does not have observable property | has observable property | has observable property | does not have observable property |
|---|---|---|---|

| contrasting properties | shared properties | contrasting properties |
|---|---|---|
| electric charge | like repels like | magnetic poles |
| charge is unipole | unlike attracts unlike | magnets are dipoles |
| moving charge makes magnetic field | force extends over distance | moving magnetic field makes electric charges flow |

## Conflict Analyzer

The Compare and Contrast Analyzer can be modified to a Conflict Analyzer to help discover the causes of disputes and conflicts. This is an important troubleshooting technique in everything from computer programming and car troubles to medical diagnosis and marriage counseling! In this Conflict Analyzer, a Venn-type diagram is combined with the Sorting Analyzer. The areas of overlap are areas of conflict. It is the observable that causes a conflicting view of the same system by two different observers.

## Cross-Curricular Example

Here is an example that relates to both current events and to disputes in the classroom. Both John and Jim perceive that a particular space is theirs. The Conflict Analyzer will show the conflict to be in John's and Jim's choice of observable.

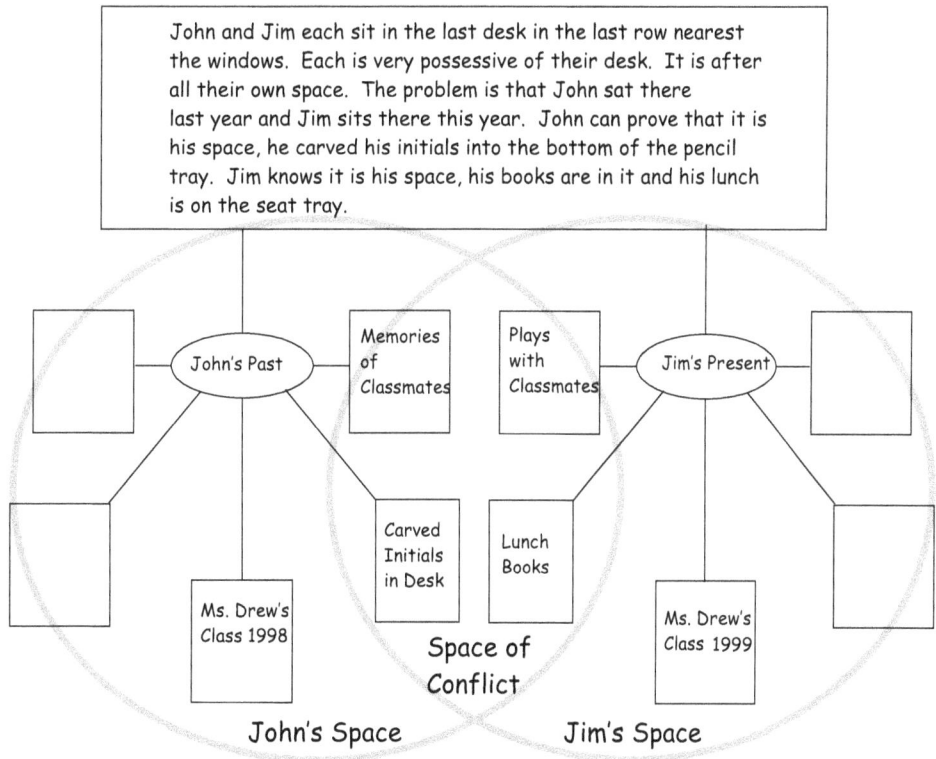

John and Jim each sit in the last desk in the last row nearest the windows. Each is very possessive of their desk. It is after all their own space. The problem is that John sat there last year and Jim sits there this year. John can prove that it is his space, he carved his initials into the bottom of the pencil tray. Jim knows it is his space, his books are in it and his lunch is on the seat tray.

John's Past — Memories of Classmates — Plays with Classmates — Jim's Present

Carved Initials in Desk — Lunch Books

Ms. Drew's Class 1998

Ms. Drew's Class 1999

Space of Conflict

John's Space

Jim's Space

# CHAPTER 23

# The Road to Understanding

This exploration of the Progression of Inquiry and Systems Analysis using the Language of Patterns and the Graphical Analyzer System is now complete. Its purpose has been to provide the tools needed to move you onto the road to the Age of Understanding.

There are many more applications and a whole world that can be explored, described, measured, and known through the Language of Patterns and the Graphical Analyzer System. The tour is ended. Your journey begins.

# The Neuroscience and Neurology of the Human Brain

## A More Detailed Exposition About the Brain

The human brain has evolved over time to become a system of efficient computing centers that acquire pattern data from the environment.  At multiple levels, elements of the patterns are abstracted by nerve or neuronal circuits and passed on to higher brain centers for analysis.

For example, when we see a face, the edges and contrasts of the hair pattern and the shape of the jaw play a major role in our assignment of an identity to whose face is being regarded.  It is for this reason that changes in hair cuts, growing beards, or masking portions of the face can make recognition difficult.

Over most of evolutionary time, the brain of the human species, *homo sapiens sapiens,* has been molded from the non-human brain. The abstraction processes of the brain have been driven by the need to ensure survival of non-human animals, and therefore many of the higher brain functions that we associate with human behavior must learn to compensate for the data discarded or manipulated by more primitive brain structures.  A knowledge of brain organization and function will help us understand why certain areas of scientific inquiry need a particular structure to be taught successfully

Patterns are the arrangement of certain elements, each with defined or definable properties in some region of a space.  The brain is designed to search for patterns and then lock onto a particular pattern. If the brain locks onto an incorrect pattern, it is not generally an easy matter to self-correct the error.

Two excellent examples of the difficulty with reversing the locking in of an error come quickly to mind.

- One is the difficulty that people have in untwisting their tongue once an improper set of motor commands is given while saying a tongue twister. In this case of a pattern of motor activity, the game is to induce the brain to lock into an improper set of command elements and ordering and then to experience how difficult it is to correct such an error without completely stopping the activity and essentially starting over again.

- The second is an experience most dog owners have had in which their dog sees a tree stump or a shadow of a garbage can, usually at night, and locks onto it as something to be cautiously regarded. No amount of reassurance, attempts to shine a flashlight, or show the dog the benign qualities of the object will get the animal to drop its fright/fight posture.

At higher intellectual centers, the human brain abstracts physical attributes of a natural pattern and then searches for patterns of <u>cause and effect</u>. It is a character of the human brain to search for and attribute causal links to the linked conditions that it detects.

Searching for causality, that is the reason why this pattern happened, is a uniquely human trait. It is a higher order synthesis than the primitive function of conditioned learning that associates events or pattern elements together. Although often confused, becoming conditioned to expect a link between elements is quite different from a true cause-and-effect relationship.

Exploring a space and finding the elements, properties, rules for repetition, and recognizing the space or background in which the pattern is found is a process that is improved and refined by learning. It is an ideal place for educational interventions. A method for correcting the error of incorrect locking onto a particular pattern can be modified by critically examining what patterns are in terms of their observed properties, rules of occurrence, and the linkages or effects that one element's properties has on another's behavior.

An extremely important aspect of the way that the brain seeks patterns and relationships lies in the knowledge that the brain is designed to search for, and find, patterns and relationships. There has been substantial evolutionary reward for this design. Unlike a digital computer that indicates that it can not complete the computation without enough data, when the brain can not find enough data in the external environment to complete the pattern-causality process, it completes the process from internal sources (often without careful differentiation of the source!).

The internal source of data can be memory, fantasy, imagination, or any form of speculation. A key function of the brain is the completion of pattern structures in the mind by <u>manufacturing</u> data when needed. This phenomena is very real and can be seen in the hallucinations generated by sensory-deprivation tanks and environments and in certain fascinating neurological syndromes such as the Charles Bonnet syndrome.

- The *Charles Bonnet syndrome* is seen in individuals in a nursing home who might have partial loss of sight due to eye disease or brain disease from a stroke, resulting in visual loss in a significant portion of their visual filled. If they are then further sensory deprived by unfortunately being placed so that the remaining visual perception is of a blank wall or darkened space, the brain responds to this sensory deprivation typically by filling in the empty sensory space with benign memories, such as the former spouse or visiting children and even cuddly rabbits, dogs, and cats. The hallucinations are usually taken as signs of a dementing process by medical workers and the family, yet the individual can be instantly "cured" by the simple re-orientation of the working eye toward a door or window with real sensory information for sampling. The hallucinations disappear and the person becomes normal again!

This physiological process of "pattern-completion" also helps explain the well documented experiences of people who come to believe that unremembered events are completely true after hearing the events told. Once incorporated into a pattern, the brain usually is incapable of distinguishing imagined elements from events that actually happened.

From an educational perspective, the absence of information relating to an experience or concept will lead the brain to complete the concept and make an attribution of causality based on the nearest and most convenient linkage. Due to the brain's structural and operational processes, the easiest and most intuitive attributions are often in a magical or primitive thinking pattern. Thus, the naive learner who has not been given proper tools to be skeptical of the natural and magical tendency for magical attribution will be prone to mislearn concepts and be unable to easily correct the errors without great and often futile effort.

## The Human Brain is a Pattern Computer

The human brain has evolved to be a powerhouse computing machine that searches out, recognizes, and organizes information into patterns. Our senses of sight, taste, smell, hearing, and touch are designed to extract specific information from the environment and present the extracted elements to various brain centers for interpretation. The interpretation of the extracted information (such as edges, light-dark contrast, and motion against a background) gives rise first to perception. As the information is passed to higher brain centers, the perceived information is interpreted and understood, giving rise to cognition. At the highest levels, information originally abstracted as sense elements (edges, contrast, etc.) is mixed in associative centers in the brain with other sensory information, stored memories, other symbolic forms, such as words or images.

To emphasize, the human brain is organized into many centers that are designed to extract certain features of the information presented to each center and then to pass on the newly abstracted pattern to the next level or to an associative area for interpretation and action. Essential to our purpose is an appreciation of how the organization and structure of the brain gives rise to the capacities and behaviors of the mind.

## A Simple Tour of a Complex Machine

The nervous system is a complex interconnected set of nerve cells designed to collect and process information and then to respond to the environment with action through reflex, motion, and communication. The human nervous system can be split into two basic units, the peripheral and the central nervous systems.

**The peripheral nervous system (PNS)** is a vast and branching array of large and small nerves:

- **Sensory nerves** carrying temperature, pressure, pain, and position information from the body inward toward the brain via the spinal cord; and

- A huge collection of **motor nerves** carrying commands telling the muscles what to do from the brain through the spinal cord.

- Autonomic nerves that control heart rate, blood pressure, sweating, skin temperature, gastrointestinal and sexual function and control of the bladder and bowels.

The **central nervous system (CNS)**, composed of the:

- the **brain (B)** which is a centralized collection of billions and billions of nerve cells all intricately interconnected supported in a jelly-like matrix of trillions of support cells that feed, nourish, and protect the brain cells...all protected inside the hard bony skull.

- **spinal cord (SC)**, a huge central trunk of nerves carrying information from the body to the brain and commands from the brain to the complicated machine that is the human body;

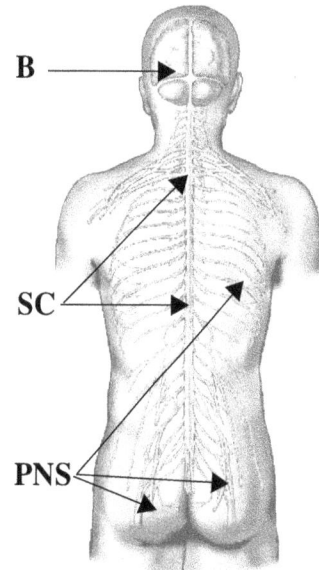

**FIGURE 62.** Central (B) and peripheral systems (PNS) are interfaced at the spinal cord (SC).

## The Human Brain is at the Apex of Neurological Evolution

If we construct an evolutionary ladder with the human brain at the advanced end and the primitive nervous system of the sponge at the other, the breadth of computing power in biological organisms is startling. There is a vast and progressive increase in complexity of the nervous system. The nervous system of the sponge is simply composed of a network of nerve cells with no specialized structures for collecting, transmitting, or processing information. The first organized structures are a very rudimentary collection of nerve cells, called **ganglia**, which act as a sort of central processing unit; these ganglia occur in the primitive worms. From worms to insects to invertebrates (such as crayfish), there are increasing numbers of nerve cells included in these ganglia, usually located at the head and tail portions of the animal. At this evolutionary level, there is a simple primitive communication pathway (a rudimentary spinal cord) connecting the ganglia with the sensory organs and muscles. The grasshopper "brain" contains no more than several thousand nerve cells compared to the 10 billion neurons in a human brain.

As the evolutionary ladder is climbed, the brain begins to accumulate another series of structures that are found first in fish, amphibians, then reptiles, birds, and finally mammals; this is the **spinal cord**. This pattern of evolutionary development is called **phylogenetic development**.

# Building the Brain

FIGURE 63. Structures in the human brain

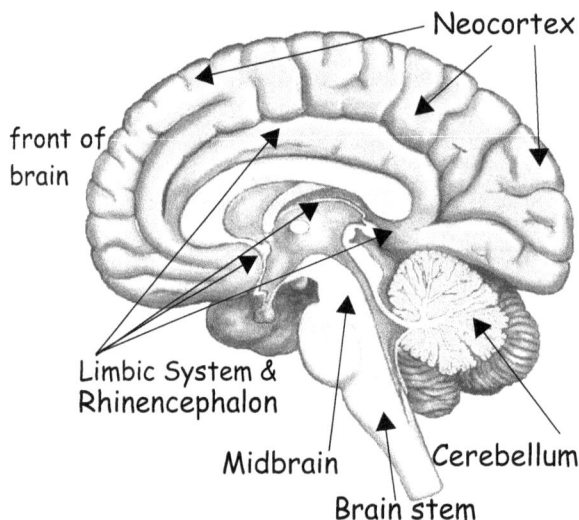

front of brain

Neocortex

Limbic System & Rhinencephalon

Midbrain        Cerebellum

Brain stem

At the junction where the spinal cord meets the brain, the collection of cells that are necessary for the most basic functions of life (control of breathing, heart rate, and digestion) are assembled into a structure called the **brain stem**. All vertebrates from fish and frogs to humans have a brain stem that serves similar functions. Sitting atop the brainstem is another collection of nerve cells the **cerebellum**.

The cerebellum is specialized to smooth, coordinate, and aid in the learning of motor activities. The frog brain has a brain stem, cerebellum, and a very small bit of brain called the **cerebrum** or **neocortex** which will develop into more sophisticated and specialized collections of cells in higher animals that will handle all of the higher brain and mind functions. The brainstem is connected to the cerebrum through a portion of the brain called the **mid-brain**

## The Nose-Brain: Sex, Satiety and Learning

The most primitive portion of the cortex is the olfactory cortex, which is a portion of the **rhinencephalon** ("nose-brain"). The rhinencephalon is well developed in fish, frogs, and reptiles and is closely associated with the specialized collections of nerve cells that are responsible for modulating hunger, satiety, sexual drive, learning, and strong emotion. Thus, the powerful influence of smell and odor is tied to intense sexual behavior, feeding, fear, and anger. The ability to learn and to form and retrieve memories requires emotional input. This is the neurological reason for the observation that highly charged emotionally events are vividly remembered. The tight interrelationship between the rhinencephalon and the nearby brain centers that control hormonal secretion also makes clear the powerful influence of the olfactory sex stimulants (pheromones) on male animals when female animals go into heat.

Irritability, or the ability to excite a nervous impulse in the nerve cell with stimulation, is the most general feature of every nervous system from the sponge to the human. If one irritable cell is connected to a second nerve cell, the irritability can be transmitted - a communication system is established. If the target of an excited cell is a muscle cell, the irritable impulse can cause movement - this is how a reflex works. Irritability of this type is the most primitive response of the nervous system.

## Learning is a Primitive Function

The ability of the nervous system to learn is perhaps the second most primitive response. This is surprising because we usually think of learning and memory as a higher function of the nervous system. We attribute both machines and animals with "intelligence" because they have memory. While memory is necessary for intelligent behavior, the formation of memory and the conditioning of neuronal circuits to alter their behavior is extremely basic and primitive. Snails and worms are capable of learning relatively complex tasks, in a manner similar to your TV and DVR.

Learning is an associative function of the nervous system. The process of learning links one event, or the memory trace of an event or object, to another object. In learning (or what psychologists call the conditioning), one event or occurrence is usually linked to another event or series of events or objects, because they occur in relative proximity to one another. In the natural world, the conditioning link does not typically relate cause to effect, but very often the human brain makes a causal attribution anyway and links the events together as cause and effect.

Because learning is such a primitive function of the brain, it resides in the oldest portion of the cortex - the **rhinencephalon**. Primitive learning is focused on associating appropriate patterns of sensory information with the needs for feeding, procreation, defense of home range, and self-preservation. In the human, the rhinencephalon has been integrated into the **limbic** system, which is a portion of the temporal lobe. As we will see shortly, the **temporal lobe** plays an important role in associating visual, touch, and auditory information with the emotional content derived from the limbic system. Thus, the temporal lobe is a central player in the association, formation, and storage of memory and learned information.

## The Neocortex

FIGURE 64. The neocortex is labelled as noted in the text

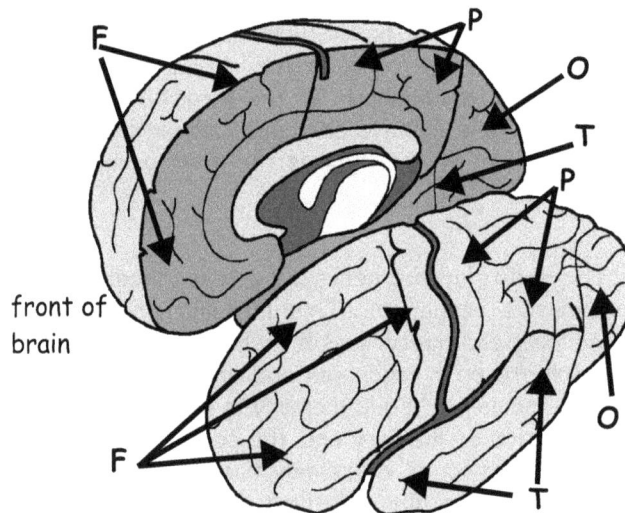

As the phylogenetic ladder is mounted, the cortex develops increasing in relative size and complexity; this newer cortex is the **neocortex**. Added levels of function extend the neurological capacities of the brain systems. In addition to the olfactory cortex, the major regions of the cortex are the **occipital lobe (O))** (responsible for visual processing); the **parietal lobe (P)** (responsible for processing and manipulating sensory information); the **temporal lobe (T)** (involved in emotional association, religious feeling, and many aspects of learning and memory recall); and the **frontal lobe (F)** (responsible for abstract thought, motor planning, speech production and planning, complex planning in a temporal plane including dimensional search paradigms, inhibition of impulsive behaviors, and delay of instant gratification).

We can explore the functions of the divisions of the neocortex by considering the role each plays in the life of the brain.

## Vision and the Occipital Lobe

First, we will explore the neurology of visual perception, cognition, and behavior. Perception is the acquisition of sensory information from the environment by the nervous system. The human eye senses patterns of light, dark, color, contrast, and motion relative to the eye or the background projected by the lens of the eye onto the retina. At a level as early as the light-sensitive part of the eye - the **retina** - the nerve cells in the retina are inter-connected, such that only certain features of the entire visual scene are selected and sent on to the visual centers in the brain. This pre-processed pattern of light is sent by way of a thick nerve bundle called the **optic nerve** to a visual center in the mid-brain. This visual center is connected to the balance centers and motor centers near the cerebellum. The visual patterns are abstracted and quickly integrated at this level to provide the type of information that allows birds and reptiles to respond quickly to changes in their environment. In humans and other mammals, the visual information is passed along to a collection of nerve cells found in the sensory way-station of the brain - the **thalamus**. Here, much of the process of abstracting the basic elements of light-dark, contrast, lines, and spots is completed before this pattern information is passed on to the visual centers in the occipital lobe of the cortex.

In the **occipital lobe**, further processing of the visual pattern elements is performed. Specific brain circuits search the visual input for curves, intersecting lines, and motion of objects against a fixed background. Color, hue, and intensity are also processed. These feature-abstracted patterns are then passed into portions of the occipital cortex and to neuronal circuitry in the temporal and parietal lobes for association with stored memory traces, auditory, body sense, and motor or movement information for planned and expected movements. There is a broad distribution of the visual information to different parts of the brain where the information is then used to form memories, learn new motor programs, associate vision with words and symbols, modify current actions, etc. There is a rich interconnection and cross-talk between many portions of the brain, as each specific center examines, modulates, and feeds- back to other areas. While the rules governing the abstraction of pattern features by the visual centers from the retina to the occipital lobe are relatively rigid, once the visual pattern is available for cortical association, an incredibly rich set of nuances is available to the brain.

## The Parietal Lobe

The important role of the **parietal lobe** can be appreciated by considering the following question.

- What is required from the brain of an animal capable of independent movement who needs to leave the den, roam about in search of food, water, or a mate, and who then returns to the safe haven of its home range?

This animal must sample many aspects of its internal and external environment in order to function in this situation. The senses supplying vision, odor, taste, touch, and memory must all be available and operational. The rich array of sensory information reaching the brain must be perceived and the patterns of data abstracted. This animal will need to carry some representation of the three-dimensional space in which it lives. Having manufactured a representation of the space in which it lives, it must be able to place itself accurately inside that space. This is a major task of the parietal lobe, which processes the complex sensory information from the external and internal environment of the animal and creates a three-dimensional virtual space that lets the animal know where it is relative to the real space in which it lives. The parietal lobe is a three-dimensional pattern computer and virtual reality machine. When the parietal lobe is damaged, as happens in strokes and Alzheimer's dementia, the animal can become lost, because it can no longer know where it is relative to its current three-dimensional space.

Emphasis is being made of the parietal lobe's three-dimensional capacity. Three dimensions are sufficient to know where the den is, how to find the river for food and drink, and to find the location of the member of the opposite sex ready to engage in mating activity. Three dimensions is what the parietal lobe can deliver.

However, the space in which we really live is not three-dimensional, it is four-dimensional! The fourth dimension is time, and our 3-D space is constantly moving forward in time. This sounds science fiction-like, but consider a classroom; it is in a specific location in a school, and if you have a class at 9 AM you must be in that specific three-dimensional location at the specific fourth-dimensional coordinate of 9 AM. The right place (3-D) at the wrong time (4-D) is not useful in the real space in which we all live. So, we are four-dimensional creatures. Before we go on, consider this question: which sense measures time? While you as an adult appreciates time; do your children, does a dog or a cat? What is the difference between adult humans and children and other animals?

## Planning Requires Four Dimensions

The answer is planning. Neither children, pets, cats, elephants, or dolphins can plan. Planning requires four dimensions, and the parietal lobe only does three dimensions. The four-dimensional virtual representation needed to plan is done by the **frontal lobe** in concert with the parietal lobe and the hippocampus.

Don't animals plan? What about the life cycle, or hunting - don't these events happen in time?

While these events certainly have a time dimension, this is an example of how knowledge of the background space allows for a better representation of an event, and therefore knowledge, but awareness of the true dimensionality of the background is not required to participate in the pattern! Consider our classroom from earlier. The people in the 3-D space of the 7 PM class need to have no awareness of the earlier classes, but the building manager could not plan the building if she didn't know about the 4-D space of the classroom.

Even though our natural world is four full dimensions, most activities and operations can be performed with an awareness and processing of only three-dimensions. Consider the sexual cycle of an animal. The timing of the pattern of the seasons (light changes mostly) activate the sex hormones. When the female is prepared to accept the sperm of the male, she secretes a strong sexual attractant, which the male of the species senses through its nose, activates its rhinencephalon, and drives it to search and find the estrous female. Mating occurs without any consideration of the impact on the animals or their environment of the consequences of the ensuing pregnancy. The stimuli and response are all in the present time. The only issue is finding the female in three-dimensional space.

Hunger is handled the same way. The animal is hungry, so it hunts. A behavior may be modulated by environmental factors: it is bright sun outside and the animal has learned that being out in the sun is dangerous, so danger is able to suppress the need to hunt for food. This is a competition of needs and wants in the present; there is no planned behavior that recognizes the fourth or temporal dimension. It is important to recognize that learning is not planning. Learning that sun and danger are linked is a conditioning response that operates in the present. In our current example - It is sunny, therefore, sun equates with danger. If the danger is greater than the hunger, do not hunt. - Later, when the sun goes down the conditioning paradigm becomes:

*It is dark. Darkness equates with no danger. If hunger is greater than anxiety, then hunt.*

We may observe this action and attribute the learned response to "planning", but this perception results from an anthropomorphism in which we unintentionally impose our own four-dimensional causality relationship onto the situation.

## The Temporal Lobe

The visual and body senses of temperature, pressure, pain, and vibration are extracted and represented in the occipital and parietal lobes. This information is in the form of "in-the-present" - three-dimensional construction. The cortex of the temporal lobe is the site of association of this sensory information with olfactory, internal monitoring (hunger, thirst, fear, satiety, anger, etc.), and emotional content information generated in the **limbic** region. As these complex patterns are mixed and associated with a level of "importance", they become available for the production of memory, which occurs deep in the **hippocampal** region of the temporal lobe. The temporal lobe is intimately involved with the execution of the learning process. This includes a complex association of multi-modal abstracted patterns, tagging the newly learned responses with emotional content, and the placing of the event in the context of the moment with the ordinal reference at the center being the self.

There is a rich interconnection between the associative cortex of the temporal lobe and the frontal lobe in humans. These interconnections are crucial for the brain to achieve the level of dimensional virtual construction that leads to planning and control of the "in-the-present" needs of the selfishly-oriented temporal-parietal-occipital-brainstem complex.

## The Temporal Lobe and Religion

A final very interesting role for the temporal lobe can be learned from observing certain neurological patients with epilepsy. Epilepsy is a condition in which certain areas of the brain overproduce nerve activity which, while it lasts, interferes with the normal overall functioning of the brain. The behaviors elicited by epileptic discharges have taught neurologists some very interesting concepts about the brain. One of the most interesting types of epilepsy involves the temporal lobe. Excessive stimulation of the neurons in the temporal lobe leads to experiences usually associated with mystical and religious feeling. For example, out-of-body experiences, deja-vu, jamais-vu, feelings of ectasy and revelation, auras of sulfurous odors, and experiences of terror and deep dread are all common during the temporal activity. In addition, the deep temporal lobe is very sensitive to a lack of oxygen or blood flow, and when this condition is imposed on people, they relate the now classic, near-death-out-of-body-rising-toward-the-light experience chronicled in literature and revelatory accounts. These findings have lead to reasoned speculation that the portion of the brain responsible for religious and mystical feeling resides in the temporal lobe.

## Working in the Fourth Dimension

The capacity to function in three-dimensional space in a fashion described earlier requires only a parietal lobe (to create a virtual 3-D space) and the temporal lobe to help make the decision based on present conditions and learned situations. There is no planning and no need for a sense of how the individual is moving through time and space. The sense organs respond to place and position and, therefore, are three-dimensional sensors. There is no sense organ that measures time. Yet, to plan an event that will occur in a particular space in the present but have an effect at some future time requires a knowledge and awareness of time. If we can not sense the passage of time with our eyes or ears or skin, how do we place ourselves in a four dimensional virtual space that allows us to plan events that take into account time?

## The Frontal Lobe Is A Time Machine

Many events do take place over time (day-night cycles, gestation periods, growing seasons, conversations) and what is required for a sense of time is <u>abstraction of the pattern element that represents the movement of the three-dimensional space through time</u> (think about your daily schedule or routine). This abstract representation of a functional relationship, that is, one that is not directly experienced, is the job of the frontal lobe.

Higher levels of abstraction that operate above the directly sensed plane of experience require a higher degree of abstract manipulation than found in the parietal or temporal lobe pattern computers. The frontal lobe is capable of manufacturing a virtual space of four or higher dimensions and then placing the individual in this higher dimension virtual space.

- The richness of how different the world looks through the frontal lobe may be best appreciated in the art of a painter like *Monet*. The impressionist artists asked many important questions about which arrangements of pattern elements and rules were necessary to represent three-dimensional space, but Monet's series of paintings exploring how the changing light of day altered the appearance and perception of the French cathedrals is an experiment in four-dimensional representation. This is a good example of how art and scientific explorations may use different tools but are often investigating topics of similar interest.

There is an advantage to this type of higher-than-3-D virtual space. Higher dimensional space provides more ways of looking at a situation. The dual-Nobel prize winner, Szent-Gorgy, summarized this capacity well, *"Genius is looking at the familiar and seeing it in new and unique ways."* Consider the differences in how a room is perceived in one dimension (to see what vision restricted to a one-dimensional line would be, look through a straw) versus two dimensions (squash everything into a single plane such as the floor) versus three dimensions. Since it is the frontal lobe that is designed to abstract space to dimensions higher than the three dimensions of the parietal animal, the frontal lobe provides the human with the ability to see into the future and to relate the past and present to the foreseeable future. ***The frontal lobe is a time machine!***

The human brain is distinguished by its large and highly developed frontal lobe. In addition to the having the ability to develop highly abstract and symbolic thought patterns (that is, explore higher than three-dimensional space), the frontal lobe is empowered to use this knowledge to control many of the other areas of the brain, such as the rhinencephalon, by being strongly inhibitory. Therefore, the frontal lobe is capable of suppressing behaviors that are impulsive and unplanned. The ability to abstract on a complex plane, to plan, and to suppress impulsive behaviors are all connected to the type of processing that the frontal lobe performs. Everyone is familiar with adults with their frontal lobes "turned off" - alcohol is a frontal lobe depressant. Becoming drunk turns off the frontal lobe, leaving a chemically lobotomized individual. The loss of judgement, enhanced aggressiveness, sexuality, and appetite seen in the alcohol- intoxicated human are all examples of the typical behavior of animals, who have no frontal lobes.

# The Neurology of Language

With little effort we could fill volumes with the wonder of the information processing capacities of the human brain. However, we will restrict ourselves here to exploring some of the neurological aspects of language and mathematics. Up to this point we have talked about the brain as if it were a single object with specialized sections added from back to front. However, one of the remarkable facts about the brain is that a plane of symmetry cuts the brain into a left and right mirror image. So, we usually talk about the left brain and the right brain almost as if they are two twins living in the same skull. As the nervous system develops, the nerve fibers entering and exiting the left brain end up connecting to the right side of the body and the right brain connects to the left side of the body. The left and right brain usually know what is happening on both sides because there are several very thick sets of interconnecting fibers talking back and forth. The parallel portions of the brain do generally the same thing, e.g., right hand movement is controlled by the nerves at the bottom of the left motor strip (found in the left frontal lobe), while left hand movement is controlled by the nerves at the bottom of the right motor strip (found in the right frontal lobe). But some functions are located more on one side than the other, this is called **lateralization**.

Language is lateralized. This makes great sense, since the left hand can still function even if it does not know what the right hand is doing; but if the left mouth doesn't know what the right mouth is doing, no communication can occur. The production of language, both written and verbal, is accomplished in the "dominant hemisphere". The great majority of people are naturally right-handed, and speech production is located in the left hemisphere. However, approximately 50% of people who are left-handed still produce language in the left hemisphere of the brain.

Language production requires the integration of the whole brain. Many aspects of language are refined and learned on exposure to language in the environment, but the process of acquiring a knowledge of patterns of language, its elements, rules, and its contextual background space is largely biologically determined. The brain has evolved to accomplish the task. Much of the dominant lobe is dedicated to the implementation of these rules.

The daily business of language can be usefully separated into two domains: 1) reception - or sensory language acquisition, and 2) production - or motor language production.

## Language Reception

If the language to be considered is auditory, then the sensory input is routed from the ears via a series of nerve centers, up the brainstem to the mid-brain (very near the mid-brain visual centers) and into a portion of the parietal lobe just at the junction of the temporal lobe. This very important area of cortex is named **Wernicke's area** and is the central reception area for words and language elements.

**FIGURE 65.** Language areas in the human brain

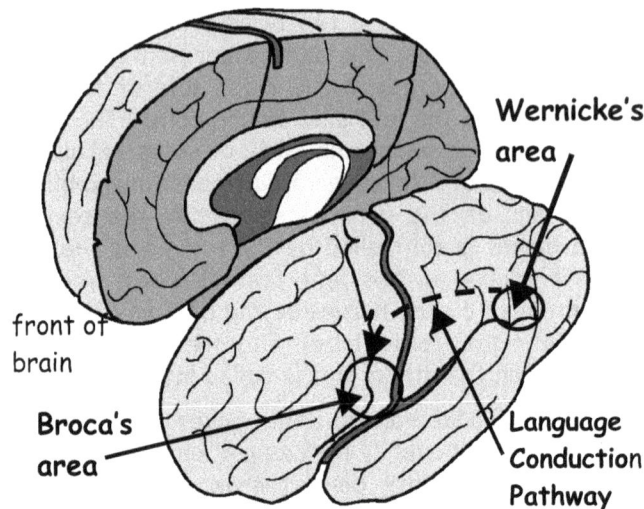

Wernicke's area is located in the dominant hemisphere, and all auditory information coming from both ears is routed to this one dominant center. Words need definitions to become meaningful, and the memory traces of visual, auditory, sensory, and emotional experience are assembled from their storage areas that are found in the associative cortical areas in the parietal, temporal, and occipital lobes. Words that are read or visually input are also formed and given meaning in this region after both occipital lobes have extracted the pattern of the word; then they are packaged and the interpreted pattern is shipped to the region of the Wernicke's area.

## Language Production

The production of language requires motor planning and coordination of the next word to be uttered with the last word spoken or heard. The region for planning the motor program needed to move the diaphragm, vocal cords, tongue, and mouth is in the frontal lobe at the bottom of the motor strip. This region is called **Broca's area**. Broca's area makes connections with motor command cells on both sides of the brain, so that the mouth moves in a coordinated and cogent pattern to make words. If the language is to be written, Broca's area sends the language output plan to the portion of the motor strip responsible for planning hand and finger movement. Under the control of the Broca's language area, the hand motor strip then executes the necessary motor program for word production by writing or typing.

There is a conduction pathway that interconnects Wernicke's area with Broca's area that must be intact for effective speech production. This conduction pathway is the site of many interconnections between brain centers from all over the dominant and non-dominant brain. Damage to any of these areas from tumor, strokes, or trauma leads to very characteristic language disorders called **aphasias**. Damage to Broca's area interferes with all language output whether spoken or written. Language production is a multi-dimensional process requiring context in the present, association with past knowledge and memory, and planning both in listening and comprehension but also in formulating and responding by speech or writing.

If the dominant lobe is busy with comprehending and producing words, does the non-dominant lobe have any role in language? Interestingly, the non-dominant lobe has a parallel role to perceive and produce, not the words but the rhythm and nuance of language - the prosody of speech.    The non-dominant temporo-parietal region detects the emotionality of speech, its gruffness, edge, sweetness, or melody. The non-dominant motor center adds modulatory tone to the speech produced. Damage to either of these non-dominant areas leads the individual to either lose the ability to sense the emotional melody of speech or destroys the ability to make words sing or bite. Musicality seems to be associated with these non-dominant regions.

## The Neurology of Mathematics

Mathematical and nominal, numerative and geometric skills are also dispersed among the structures of the brain. Just as the brain seems to be biologically wired to extract the patterns and rules of language from spoken language, the same intrinsic pattern recognition that allows nominal ordering of patterns and its associated sense of numerativity is un-critical but innate.

## Nominal Ordering Resides in the Dominant Parietal Lobe

The ability to manipulate sets of numbers to count arithmetically, e.g., three apples and two apples equals five apples, is a skill that resides in the dominant **parietal lobe**. In mathematical terms, arithmetic of this type is just a function of pattern manipulation using nominal ordering of elements in sets. It is interesting that the dominant parietal lobe is also the location for detailed property assignment to items. For example, naming the fingers and knowing left from right, both of which are clearly nominal orderings, are functions that reside near to the areas for arithmetic calculation in the dominant parietal lobe. Finally, a high degree of  detail used to describe an object also is a nominal characteristic of a pattern element and is neurologically assigned in the dominant  parietal lobe.

## Three Dimensional Construction Is A Characteristic Of The Non-dominant Parietal Lobe

While the dominant parietal lobe is busy attending to the details of name, number, and details of properties, the non-dominant side constructs the three-dimensional space in which those detailed pattern elements can reside. If asked to draw a house, an adult with an intact non-dominant parietal lobe but damaged dominant parietal lobe will draw a perspective or three-dimensional representation of the building but leave out the details in the picture. The ability to draw a perspective cube requires frontal lobe planning, and usually the ability to construct such a shape doesn't develop until after 12 years of age; it requires an educational level of about the 6th grade level. Alternatively, damage to the non-dominant lobe alone will cause a house of flat dimension with ample detail to be generated. Damage to the non-dominant parietal lobe can lead to a lack of recognition of where a person's own body begins and ends, a person may regard their own left arm as belonging to someone else.

## Biology and Behavior

Human development and disease has provided a great deal of information on how the brain functions and gives rise to the behaviors that we associate with the mind. At birth or soon thereafter, virtually all of the specific brain centers are present and functioning, but these centers do not become fully interconnected until early adulthood. This is a result of the biological fact that the nerve fibers that will carry the information are not covered in the insulator material called **myelin** until varying times in the first 20 years of life. Thus, the information being carried by the nerves is short-circuited and is not usable. In the brain, the most primitive regions become interconnected first, followed in various order by the most recently evolved structures. The order of connection of the human brain follows the pattern of evolution of the brain itself, thus **ontogeny**, or the development of the individual brain, follows the **phylogeny**, or pattern of the evolution of the brain in general.

The pattern of myelination in the developing human explains many of the observed milestones in a child's development. Several examples are helpful:

- A primitive reflex called the *grasp reflex* causes the toes of the foot or the fingers of the hand to wrap tightly around whatever object touches the palm or sole. This reflex is probably very useful if the child is swinging through the trees, but unless the reflex is suppressed, the hand can not be used for fine motions that require grabbing and letting go. Also, if the foot grasps at the ground, walking (which requires repetitive placement of the foot and then release with subsequent replacement) is impossible. Inhibitory nerves are in place to suppress these reflexes but are unmyelinated until about 6 months of age for the hand and about 12 months for the feet. The age at which a child begins to gain some dexterity with the fingers is around six months and is able to start to toddle is about 12 months, when these nerve fibers eliminate.

• When the bowel and bladder are filled, these organs will involuntarily contract and evacuate themselves. Toilet training requires that bands of voluntary muscle (sphincters) that prevent the involuntary emptying become functional and under the control of the brain. The pattern of myelination is such that only between 18 and 24 months of age do the nerves that connect these voluntary sphincters to the brain become functional. The timing of toilet training is defined to a large extent by the biology of the myelination pattern.

## Concrete Thinking

Does this type of biology explain aspects of child behavior and learning patterns? Children up through late adolescence tend to be concrete, learning each specific pattern, activity, or fact without a tendency to a broader generalization of an underlying concept. They tend to focus on specific properties and to resist generalization of properties to other objects; they have a very poor sense of time and plan poorly. These tendencies are the **concrete thinking patterns**.

The concrete behavior of children is easily appreciated in the context of myelination patterns. The structures of the brain responsible for extraction of visual, auditory, and tactile information are intact and functioning in early childhood. These phylogenetically older portions of the brain provide for efficient extraction of sensory information, construction of three-dimensional space, awareness of hunger, thirst, discomfort, and other more basic wants and needs. The portion of the brain responsible for fear and glee, magical thinking, and the anger of a tantrum are all resident in the limbic area and neighboring regions of the brain. The child's brain is essentially just the brainstem, midbrain, occipital, parietal, and temporal lobes. However, the newest and most uniquely human portion of the phylogenetic brain, the frontal lobe is relatively unconnected to these more primitive areas in the child's brain. The connections develop throughout life but are complete only in early adulthood. It is the frontal lobe that is responsible for the unique sense of time that leads to the adult-like traits of planning, delaying gratification, and logical-rational abstract thought. Education can enhance these adult-like qualities, and many psychometric tests measure these adult-like tendencies (the IQ test is a good measure of frontal lobe connectedness as well as capacity). Obviously, there is a broad range of both intrinsic capacity that, coupled with patterns of myelination, can lead to varying capacities for abstraction, planning, and delayed gratification (often manifested as studiousness) in the age ranges of children.

## Social Implications Of Myelination Patterns

One of the difficulties in regarding older children and adolescents as "small adults" is that without the full capacity of the frontal lobe connected into the remainder of the brain, adult behaviors are very difficult to elicit reliably. For example, it is unreasonable to expect the average 13 or 14 year old to appreciate the implications of how being sexually active today may have dramatic consequences in nine months and that event in nine months may completely change their potential future. Since the young adolescent is not yet intellectually capable (frontal lobe competent) of moving through the virtual time dimension, expecting a "rational", that is, adult-like, analysis is unfair to ask of the child and foolish with respect to society's best interests. The same "poor" planning should be expected for many actions to which adolescents are given access when the "young adult" error is made. The inadequacy of the frontal lobe connections is especially apparent when strongly emotional or rhinencephalic actions are considered. Thus, in the areas of violent, sexual, drug and alcohol usage, and impulsive behaviors, the pre-adult is operating with a fully functioning set of brain circuits capable of the behavior but without the benefit of the frontal lobe which would modulate and moderate the behavior only if fully connected and operational.

## The Neurology of Patterns and Categories

There has been tremendous evolutionary pressure to be able to categorize and generalize effectively and accurately in a concrete, predominantly three-dimensional world. Patterns of information in the form of light, chemicals, and mechanical pulls and pushes are extracted and abstracted into a category or group:

- This round red object is safe to eat.

- This round blue object makes me sick.

- The horned four-legged animals at this water hole are good to eat and can be caught.

- The land becomes covered with cold white death after the sun only rises one hand above the mountains.

The human brain is designed to generalize objects into groups. The brain, therefore, has evolved a sophisticated pattern recognition apparatus that is fundamental in the learning of virtually all primitives used in exploring and gaining knowledge about a system. The problem with abstraction is that chosing only certain properties of an object to make it fit into a group or category reduces the resolution of what can be known about the object. However, in exchange for this loss of information, information processing is greatly speeded. Evolution has clearly walked a careful balance to build its neuronal pattern computers to be fast, efficient, and, usually, accurate. Abstracted information is passed on to other brain areas as a pattern itself, ready to be further abstracted. If the object is abstracted incorrectly, the information necessary to correct the error may have already been lost. Due to its organization, the mind tends to hold onto any concept, *including misconceptions,* tenaciously. Thus in time, if eating round objects, in general, is not required for survival, a generalization focusing on their dangers may become the only operative function; that is, round objects cause illness, avoid them.

The most basic and essential task that the brain performs is determining if a particular object or event fits into a particular category or not. This fundamental brain operation, which has deep philosophical as well as practical implications, may be quaintly restated as the "dogginess" question, i.e., how can we so easily look at any of the over one thousand species of dog and know that they are dogs and not cats, wolves, or coyotes? Humans perform the same basic operation everyday with all sorts of objects.

The categorization of objects (both physical and conceptual) into various groups, often with complex interrelationships, is a crucial step in the process of learning about new systems and monitoring and utilizing existing systems. This process is so natural that we often don't even consider it, yet clinical medicine has provided us with an "experiment of nature", that is, a group of individuals who are incapable of making easy categorizations - autistic brains. The problem that an autistic person has in relating to the world is that every object stands alone; it can not be categorized into a group and assigned group properties. Thus, every dog or face stands out individually and is not easily treated with any abstraction. There is no such thing as dog-like behavior, because everything is the behavior associated with that particular dog, or that specific dog over there, or that one over there.

Such a level of detail and concrete un-abstractable space makes it very difficult to confront the richness of the natural world. Thus, we know that the inability to abstract and categorize is disadvantageous.

## Hierarchies of Pattern Abstraction: Magical and Skeptical Thinking

Once a pattern is perceived, the brain goes on to assign casual relationships between the elements of patterns. Through most of human history, the causality for patterns has been attributed to magical forces and gods of various supernatural powers. It is particularly interesting that the centers in the brain for religious feeling, and mystical and magical emotion seem to be located in the temporal lobe, a portion of the cortex tied tightly to the emotional and older rhinencephalic brain. The temporal lobe may thus play an important role in making religious and theological attribution to events of enormous survival value, such as eating, fighting, and reproducing. The temporal lobe, like its neighbor the parietal lobe, however, are portions of cortex that manipulate a three-dimensional space that is placed concretely in the present tense. These are descriptions of a kind of child-like thinking that we call **pre-critical**. The combination of the abstraction process that is prone to discard important underlying data and a magical process of causal attribution makes this pre-critical pattern recognition prone to error.

The capacity to ask questions about the mystical attribution of cause arises as the frontal lobe engages in the pattern analysis. Because the frontal lobe is capable of higher dimensional abstraction, it provides the capacity for a much deeper sense of self-awareness and questioning of the role of the human animal in the immediate universe. If culturally released from theological dogma, the frontal lobe has the capacity to use rational analysis and logic to explore the patterns with which it works. Thus, the **critical** phase of thinking starts when the frontal lobe becomes partially operational in the overall structure of the mind. Critical thinking is a form a skeptical analysis by the frontal lobe of the concrete-magical formulations of the non-frontal brain. As we have seen in our exploration of the history of science, the higher dimensional thinking of the frontal lobe that manifests itself in patterns of logic and rational thought are also susceptible to a rigid application of pattern description that can obscure the tendency of the brain to make further errors in the exploration of the natural world. To prevent the natural tendency of the pattern recognizing circuits of the brain from making errors both in perception and in causal attribution, modern skeptical science has evolved using experimental techniques over just the last 350 years.

Ultimately, it is the frontal lobe that is responsible for giving an individual a sense of his/her place in the general sweep of time. Without it, the brain perceives the world as centered concretely around its three-dimensional self (a very child-like behavior). This concrete fixation in three dimensions is pre-critical. With fully operating centers of magic and mysticism (in the temporal lobe), the thinking of the non-frontal brain tends to be magical and non-rational. As the brain develops the capacity for more restraint and rational questioning, as in a partially connected frontal lobe, the brain enters a critical thinking phase but without the balance that comes from appreciating the actual fit of oneself into the flow of time (four-dimensional space). The balance comes with recognition that the brain itself is part of the picture and therefore, it is susceptible to perceptive, cognitive, and abstractive error. The frontal lobe itself is susceptible to error but is capable of allowing the adult brain to use experimentation as a method of reality testing to avoid the errors intrinsic to its own phylogenetic structure. In other words, the frontal lobe allows the mind to abstract itself to some degree from its three-dimensional prison and appreciate, if not sense, the full *multi*-dimensional space in which it resides.

## Ontogeny, Phylogeny and Learning Science

Thus, the ontogeny of the individual from <u>pre-critical</u> (child-like: concrete and magical) through <u>critical</u> (adolescent: rigidly-abstract) to <u>modern-empirical</u> (adult-like: *multi*-dimensional abstract with experimental error-checking) follows the phylogeny of the brain through the same stages. Historical patterns of scientific thinking have followed the same track. The role for education is fundamental, because from critical stages onward, the formalisms (language, mathematics, technical expertise, etc.) required for both philosophical extraction and error-checking function to sense higher-dimensional space are non-intuitive. These non-naive processes require formal structure, especially since many of the modern-empirical error-checking mechanisms are socially complex and require the inter-personal gauging of control in experiments, calibration in measurement, context in analysis, and communication to ensure critical assessment.

## Summary

In summary, the brain processes information by first detecting a pattern of elements that are abstracted and passed on to a higher brain center via connected paths. At each center, the abstracted elements are manipulated and often combined and compared with information from other centers. A very sophisticated treatment of the pattern elements at the level of the parietal lobe constructs a very detailed 3-D virtual space into which the brain places itself. In the adult human, the addition of the frontal lobe allows abstraction of patterns to higher dimensions, such as time, which allows for analysis that leads to planning, control of instant gratification, and abstract thinking.

For more information about
SymmetryScience™ materials


Symmetry Learning Systems, Inc

**www.symmetrylearning.com**

www.ingramcontent.com/pod-product-compliance
Lightning Source LLC
Chambersburg PA
CBHW062058090426
42741CB00015B/3271